Encyclopedia of Seas

This Encyclopedia of Seas series is designed to accumulate and systematize our knowledge about the unique natural water areas — the Aral, Caspian, Black, Arctic, Far-Eastern, and Baltic seas; their wealth; the events that took place on its waters and shores; and the remarkable people whose lives were and are closely intertwined with the seas. The Encyclopedia series contains thousands of terms and concepts related to the seas. It describes geographical features: rivers, lakes, straits, and bays; provides information about towns, seaports, transport communications, basic aquatic biological species, nature reserves, national and international programs for the study of the sea, research institutes, historical monuments, activities of prominent explorers and travelers, researchers, and scientists. Each Encyclopedia includes a chronology of major historical events connected with these seas for several centuries.

Igor S. Zonn

The Ionian Sea Encyclopedia

With 185 Figures

MOSCOW WITTE UNIVERSITY
QUI NON PROFICIT, DEFICIT

Springer

Igor S. Zonn
Engineering Research Production Center for Water
Management, Land Reclamation and Ecology "Soyuzvodproject"
Moscow, Russia

ISSN 2626-1383 ISSN 2626-1405 (electronic)
Encyclopedia of Seas
ISBN 978-3-031-08205-4 ISBN 978-3-031-08206-1 (eBook)
https://doi.org/10.1007/978-3-031-08206-1

Translation from the Russian language edition: "Ionitscheskoje More Enziklopedia" (The Ionian Sea Encyclopedia) by I.S. Zonn, A.G. Kostianoy, A.V. Semenov, A. Joksimović and M. Đurović,

This Springer imprint is published by the registered company Springer Nature Switzerland AG.
The registered company address is: Gewerbestrasse 11, 6330 Cham, Switzerland

The Ionian Sea Encyclopedia

The book is dedicated to the Ionian Sea, which is part of the Mediterranean. The encyclopedia contains about 600 entries on the hydrographic and geographic objects, hydrological features of the sea, biological resources, as well as administrative-territorial units of the Ionian countries. The most significant natural objects like islands, peninsulas, bays, rivers, and mountains and their geographical peculiarities are briefly described. The economy, culture and history, cities, ports, international agreements, research institutions, and activities of outstanding scientists, researchers, and travelers are also presented in the publication. The chronology of the main historical events that have become significant landmarks in the history of discovery and exploration and of the Ionian Sea from the eighth century AD to the present day is given.

(1) Engineering Research Production Center for Water Management, Land Reclamation and Ecology "Soyuzvodproject", Moscow, Russia
(2) S. Yu. Witte Moscow University, Moscow, Russia

Igor S. Zonn

Preface

In 2021 Publishing House "Springer" published *The Adriatic Sea Encyclopedia.*[1] It was the first encyclopedia describing one of the seas which is involved in the Mediterranean Sea.

The Ionian Sea Encyclopedia is the second book from the series The Seas of the Mediterranean Region. The Ionian Sea is the transit sea, which is situated in the north of the Mediterranean Sea, between the Adriatic Sea in the north Aegean Sea in the east, Balkan and Apennine Peninsulas, and Sicily Island Three European countries are bathed by the Ionian Sea waters: Albania in the north, Greece in the west, and Italy in the southeast; however, sometimes, it i considered that the sea is bounded only by two countries – Greece and Italy This is the deepest sea in the Mediterranean Basin. The Ionian Sea with it shores is the cradle of mythical and, actually, great history. This is where the centuries-old European civilization, seamanship, maritime culture, and trade have been conceived resulting in the interaction among peoples. This is the history of the most famous figures, from Odysseus – a hero of the Trojan Wa and the legendary king of Ithaca – to "golden" Greek Onassis, a billionaire and a shipping magnate (twentieth century). Great Archimedes and Pythagoras and George Byron, a British poet and a Greek national hero, have left thei trace in this historical mythology. Cicero, a Roman statesman and a writer, ha said about the cities scattered around the Mediterranean Sea: "Like frog around a pond we have settled down upon the shores of this sea." The shore and islands of the Ionian Sea were described in the epic poems of Homer, Iliad and Odyssey. The Greek influence was extended through sea routes. The history of this predominance is sealed in antique and medieval fortresses monasteries, places, and mythical edifices, which often became a catch fo pirates of the Ionian Sea and conquerors involving the Byzantines, the Ottomans, the French, and the British. By the beginning of the seventh century B.C., Ionian Greeks established the coastal colonies around entire Black Sea area.

Democracy was born on the shores of the Ionian Sea. F. F. Ushakov, the most illustrious Russian naval commander, laid foundations of democracy in the constitution of the Ionian Islands and founded the Seven Island Republic. Ioannis Kapodistrias (1827–1831), the Foreign Minister of Russia and the first Governor of independent Greece, also left his mark in the history

[1]Zonn IS, Kostianoy AG, Semenov AV, Joksimović A, Đurović M. The Adriatic Se

howed heroism and nobility when rescuing people of the Italian City of Messina during the catastrophic earthquake (1908). The sea is famous for he great naval battles of Lepanto (1571), Corfu (1799), and Navarino (1827).

The Ionian Sea is a sea of the freedom-loving Illyrians and Greeks, fighting for their independence through their steadfast perseverance and eternally oving the clearest sea with a violet hue of waters (seas of the Mediterranean Sea differ in color).

Nature endowed the sea and surrounding coasts with natural treasures, Mediterranean climate, numerous picturesque islands and bays, beaches with clearest sands, immensely beautiful mountains, rich Mediterranean vegetation, and diversity of marine flora and fauna. A black Kefalonia spruce, Mount Etna, he Corinth Canal, the largest cable-stayed bridge in Europe, and many others andmarks are the most famous attractions of the area.

The Ionian Islands of Greece with their unique exotic Blue Flag beaches are particularly popular among tourists from all over the world. The Ionian Islands nvolve the large islands – Lefkas, Kefalonia, Ithaka, Zakynthos (Zante), Kerkira (Corfu), Kythira – and numerous small islands. All islands and the continent are interlinked by maritime traffic including ferry service.

Here, there are excellent marinas for yachts, motorboats, and boats. You can ind a perfect chance to enjoy snorkeling, windsurfing, kite surfing, and diving. All this is accompanied by olive and pine groves, vineyards and citrus gardens, excellent cuisine and vines with a smell of the Ionian Sea, and beautiful preserved edifices, which make the water area so attractive.

The encyclopedia contains about 600 concepts and terms covering the lescription of the Ionian Sea and its coast. It characterizes geographic objects: ivers, straits, gulfs. In addition, the encyclopedia provides information about cities, seaports, transport communications, main aquatic biological species, wildlife sanctuaries, national and international programs on the study of the ea, scientific institutions, historical monuments, activity of the eminent naval commanders, researchers, and scientists.

The rather small and accessible publication of the encyclopedia is intended o meet the needs of many readers in the knowledge of the history and geography of the Ionian Sea. Anticipating possible comments on the text, we lraw your attention to the author's approach to the selection of natural-historical facts.

In our opinion, the historic significance of these facts providing an insight nto the past and the present of the Ionian Sea and the surrounding area is the nain selection criterion. Therefore, we have a responsibility for possible naccuracies and interpretations, which can be met. The challenge in selecting and writing articles was a wealth of popular touristic information, on the one hand, and multilingual presentation of this information, on the other.

It should be noted that geographical names caused serious difficulties. We vould rather not adhere to any atlas or maps, where the geographical names vere fixed, because every source contains some inaccuracies conditioned by ime of its compiling. Thus, we have been guided by more than one source, and, wherever possible, duplicated their names.

knowledges of the nations, who live on the coasts of the Ionian Sea. It is quit natural, that the data, which are represented on the encyclopedia pages, are no exhaustive. We were pleased to collaborate with scientists and professionals who have studied the Ionian Sea, in order to systematize all knowledge gaine in this field.

The project team included many researchers, which participated directly o indirectly in the creation of the encyclopedia; some of them are listed in th reference list. The use of the materials involuntarily made them our co-authors

We would like to express our gratitude to the Director of Institute of Marin Biology, Kotor, Montenegro A. Joksimovic for many years of cooperation an assistance in the work. We are particularly grateful to Denis Vladimirovicl Denisov ("Mezhdunarodnye otnosheniya" publishing house) for the searcl and selection of illustrations related to different articles of the encyclopedia.

The invaluable assistance in the preparation of the encyclopedia wa provided by Larisa Aleksandrovna Abakumova, whose benevolence an computer skills allowed us to bring the work to a successful conclusion.

We express our deep gratitude to the Rector of the S. Yu. Witte Moscov University, Aleksandr V. Semenov, for financial support of this publication.

Moscow
October 2022

Prof. Igor S. Zon

LBANIA

apital – Tirana
rea – 28,748 km²
opulation – 2.8 million
eople (2020)
atural resources – oil,
as, lignite, various ores
eclared on November 28, 1912
 member of the United
ations since 1955
n associated member of the
uropean Union

REECE

apital – Athens
rea – 131,957 km²
opulation – 10.7
illion people (2020)
atural resources – nickel,
auxites, manganese, gold
eclared on March 25, 1821
 member of the
nited Nations since 1945
 member of the

ITALY

Capital – Rome
Area – 301,000 km²
Population – 60.3 million
people (2021)
Natural resources – various ore
marble, granite, tuff
Unified on March 17, 1861
A member of the
United Nations since 1955
A member of the
European Union since 1957

Basic Abbreviations and Acronyms Used in The Ionian Sea Encyclopedia

A.D.	Anno Domini
ASL	Absolute sea level
AS USSR	Academy of Sciences of the USSR
B.C.	Before Christ
Bln	Billion
Cent.	Century
Cm	Centimeter
COE	Council of Commission
Et al.	And others
EU	European Union
FAO	Food and Agricultural Organization (United Nations)
IAEA	International Atomic Energy Agency
ICZM	Integrated Coastal Zone Management
Is.	Island
Isl(s)	Islands
Kg	Kilogram
Km	Kilometer
Lat	Latitude
Long	Longitude
Max	Maximum, maximal
Mln	Million
Mm	Millimeter
MSP	Marine spatial planning
NM	Nautical mile
NATO	North Atlantic Treaty Organization
Navy	Naval Forces
OSCE	Organization for Security and Cooperation in Europe
Pen.	Peninsula
PR	Petroleum refinery
R/V	Research vessel
RAS	Russian Academy of Sciences
SL	Sea level
T	Ton
Thsd.	Thousand
UNESCO	United Nations Educational, Scientific and Cultural Organization

A

Acarnania

Acarnania – the historic western coastal region of central Greece between the bay (Amvrakikos) in the north and the Gulf of Corinth in the south, washed by Ionian Sea. The inhabitants of the region were mainly engaged in the cultivation of grain crops and grapes. Acarnania with its open western coastline and its location in the immediate neighborhood of the islands of Ithaca, Kefalonia, and Lefkada (Lefkas or Leukas) was of great strategic importance. In 342 BC Acarnania contracted an alliance with Athens against Sparta, in 27 BC was incorporated into the Roman senate province of Achaia.

Acatium (*lat. acatium*)

Acatium (*lat. acatium*) – (1) Ancient Roman 1–2 mast vessel with lugsail rig; (2) single-masted sailing boat of pirates of antiquity; (3) a three-masted European sailing and rowing vessel of the end of the eighteenth to beginning of the nineteenth century, used for military purposes in coastal areas. Its armament consisted of 18 guns, 10 pairs of oars and lugsail rig. Eight Acatia were built for the Russian fleet at the end of the eighteenth century. The first two Acatia joined F.F. Ushakov's Mediterranean squadron. The length of the vessel is approx. 29 m, its width is approx. 8 m, its water draft 3–3.4 m. The crew contains 100–107 people.

Achelous (*Acheloos*)

Achelous (*Acheloos*) – the largest river in Greece, which flows into the Gulf of Patras 6.5 km northwest of Cape Scrofa in the extreme southwest of the Balkan Peninsula in Western Greece. The length of the river is 217 km. It originates in the Pindus range at an altitude of 2000 m. The average width of the river is 90 m, depth is 2.7–7.3 m. The basin area is 6329 km^2. In winter, it spills over the entire plain adjacent to the sea. The rise and fall of water in the river occurs unexpectedly. At the mouth of the river lies a low grassy islet, to the south and north of which there are two narrow entrances to the river (Fig. 1).

In ancient times, the river served as a natural border between Aetolia and Acarnania. In ancient times it was called Thoas, Thestius, or Axenus. It was later named after Achilles. In the 1960s dams and hydroelectric power plants of Kremasto and Katsiki were built on the river, which provide electricity to Western Greece. The river is navigable.

"Achelous" ("*Acheloos*")

"Achelous" ("*Acheloos*") – gunboat (steam barque), a ship of the Greek fleet of the late nineteenth to early twentieth centuries. One of the four "river squadrons" (q.v.), bearing the names of Greek rivers. Others were "Alfeios," "Eurotas," and "Pineios." The name "Achelous" is the name

I. S. Zonn, *The Ionian Sea Encyclopedia*, Encyclopedia of Seas,
https://doi.org/10.1007/978-3-031-08206-1

of the river. It was built in 1884 at Blackwall Shipyard, England. Length 51.5 m, width 7.6 m, height 4 m, displacement 420 t, three sail masts, 400 h.p. engine, speed 10 knots, and crew 70 people. Armament 2 × 76 mm guns. She took part in the Greco-Turkish War of 1897. In 1912 she took part in the landing of troops led by Major S. Spyromilios (q.v.) to liberate the town of Himara. In 1913, "Achelous" together with the "squadron of rivers," a number of destroyers and the flagship ironclad warship "Psara" cruised along the coast of Epirus in Albania from Preveza to Avlonas, and then to Dyrrachium. She took part in the capture of the Ioannina fortress. At the end of 1916, all four ships of the "river squadron" (q.v.) were confiscated by the Entente due to the initial neutrality of Greece and transferred to the French. However, after Greece joined the war in 1917, the ships were returned to the Greek fleet in 1918. The gunboat was used as a training ship for the midshipman's school and in 1925 was withdrawn from the fleet. Subsequently, a minesweeper (1943) and a large landing ship (1947) were named after "Achelous" (Fig. 2).

Fig. 1 Achelous River. (Source: https://xn%2D%2D%2D%2D8sbiecm6bhdx8i.xn%2D%2Dp1ai/sites/default/files/images/strani/Greece_reki_5.jpg)

Aci Castello (*Aci castello*)

Aci Castello (*Aci castello*) – a small town located in the province of Catania on Sicily Island north of the city of Catania, on the shores of the Gulf of Catania, Ionian Sea. Surrounded by picturesque settlements – Aci Catena, San Gregorio di Catania, Valverde, Italy. Acireale (q.v.) a popular resort known for its thermal springs and colorful carnival is located nearby. Agriculture and industry are the main branches of the Aci Castello economy. Aci Castello grew around the castle, which was built in 1076 by the Normans upon the foundations of a seventh-century Byzantine fortification. In 1169, after the eruption of Etna, many of the surrounding settlements became

Fig. 2 "Achelous". (Source: https://upload.wikimedia.org/wikipedia/commons/3/31/Acheloos.jpg)

uninhabitable, Aci Castello began to develop. In later times, the castle became the property of the bishops of Catania. In 1296, Ruggiero di Lauria, who commanded the Aragonese fleet during the infamous War of the Sicilian Vespers, was granted Aci Castello and the castle as a fief for his successes at the service of King Frederick III of Sicily. When relations between the two men soured and di Lauria transferred his loyalties to the Angevins, the castle was besieged and captured by King Frederick and di Lauria stripped of his fiefs. In 1320, Aci Castello and the castle passed into the ownership of Blasco II de Alagona. While he defended Palermo from the attacking Angevins, a certain Bertrando di Balzo attacked the city and sacked it. Today the castle houses a history museum.

Acireale

Acireale – a resort city located in the center of the Ionian Riviera – the "azure coast" of Sicily Island between the cities of Catania and Taormina, Italy. Population – 52,300 people. The area is famous for its volcanic scogliera beaches, beloved by the local Sicilians. There is also a natural reserve La Timpa, extending into the sea with its rocks. Some historians state that between Catania and Acireale, the cunning Odysseus and his comrades landed on the Sicilian coast. Here they met the Cyclops Polyphemus, were in captivity with the cannibal, and were freed from captivity with the help of resourceful Odysseus. Realizing that he had been deceived, the outraged Cyclops threw huge fragments of rocks at the escaping sailors. They are now recognized in the rocks protruding from the water (faraglioni) near Acitrezza (north of Catania). Acireale, however, is famous for a less fabulous legend: in 396 BC, it was here that the red-hot lava from Etna blocked the path of the Carthaginian army of Himilco. The city lies at the foot of Mount Etna at an altitude of 150 m above sea level. It is a place formed by lava flows, famous for its thermal springs. According to legend, the name Acireale comes from the name of the shepherd Achi, in whom the nymph Galatea fell in love. However, Galatea had another admirer – the Cyclops Polyphemus. He could not stand the fact that the nymph chose another, and killed Achi by throwing a huge stone at him. The inconsolable Galatea was only able to turn the blood of a young shepherd into a fast river, which became known as Acis River, Fiume di Jaci. Over time, the river disappeared underground, leaving behind only a few sources. The city of Akis is already known in the era of Ancient Rome – it is mentioned in written sources from the times of the Punic Wars (Fig. 3).

The name of the city has changed several times over the past millennia – Acium (in the Byzantine era), Al-Yaj (in the time of the Arabs), Achi Aquila (in the time of the Spaniards). In the fourteenth century, Achi Aquila was moved to its current location near the Achi fortress. In the seventeenth century the Spanish king Philip IV bestowed the title of "royal" on the city for its political and economic merits. This is how the modern name appeared – Acireale. Acireale is a popular resort, famous for its unique architecture in the Sicilian Baroque style and a colorful carnival – one of the most popular in Italy. The carnival has been held annually in late January to early February since 1594. Acireale is known as a city of arts – the oldest academy in Sicily, the "Accademia dei Dafnici e degli Zelanti," is located here. Most of the buildings in Acireale date from the eighteenth century, since all the old buildings were destroyed during the terrible earthquake of 1693. The city had to be actually rebuilt. In the very center of Acireale, there is Piazza Duomo, which is overlooked by the most significant city attractions: the Cathedral, the neo-Gothic St. Peter's Basilica, Palazzo Municipale, and Palazzo Modo. Among other monuments let us note the Collegiate Basilica di San Sebastiano of the eighteenth century, the oldest extant Basilica of Saint Anthony of Padua, the Church of San Biagio, and several magnificent palaces, including the Palazzo Pennisi. The main natural wealth of Acireale is its thermal springs. On the southern outskirts of the city are the so-called Thermal baths of Santa Venera – a neoclassical complex with an internal

Fig. 3 City Aciriale. (Source: http://www.milazzosiciliainconcerto.com/wp-content/uploads/2016/04/Acireale.jpg)

English garden. The local water is enriched with sulfur, bromine, iodine, and mineral salts. Since the days of Ancient Rome, it has been used to treat many diseases. Another thermal springs – Thermal Baths of Saint Catherine of Alexandria – are located in the small village of Santa Caterina in the vicinity of Acireale.

"ACTA ADRIATICA"

"ACTA ADRIATICA" – International Oceanographic Journal, published by the Institute of Oceanography and Fisheries in Split, Croatia, since 1932. For all the years of publication, the purpose of the journal has remained unchanged: the publication of scientific papers on various disciplines, mainly fauna and flora, which help to understand the Adriatic Sea as well as the Mediterranean Sea, since it is a part of it. The journal publishes papers written by domestic and foreign scientists. Much of the articles are published in English. Since 1979, the magazine is released twice a year.

Actium or Aktion (*Aktio*)

Actium or Aktion (*Aktio*) – is a cape known for the famous naval battle which took place here in 31 BC. In this battle the fleet of Mark Antony was defeated by the Roman emperor Augustus (q.v.). Next to it there is a canal, where during the battle the fleet of the legendary Cleopatra stood ready to come to the aid of Antony.

Adriatic Euroregion

Adriatic Euroregion – created in 2006 in Pula, Croatia, for transnational and intergovernmental cooperation between the regions of the Adriatic coastline. A.E. is an institutionalized scientific framework for jointly defining and solving important issues in the Adriatic region. A.E. consists of 26 members – regional and local authorities of Albania, Bosnia and Herzegovina, Croatia, Greece, Italy, Montenegro, and Slovenia. The purposes of A.E. are: to create an area of peace, stability and cooperation; protection of cultural heritage; protection of the environment;

sustainable economic development in particular of tourism, fishery, and agriculture; solution of transport and other infrastructure issues. A.E. was divided into six technical commissions: commission for tourism and culture; commission for fisheries; commission for transportation and infrastructure; commission for environmental protection; commission for economy; commission for welfare.

Adriatic Expedition (Russian Navy Adriatic Expedition 1805–1807)

Adriatic Expedition (Russian Navy Adriatic Expedition 1805–1807) – the transition of Russian Mediterranean squadron under the command of Vice-Admiral D.N. Senyavin from Baltic Sea to Adriatic Sea and fighting of the Russian fleet in Adriatic to defend the Black Sea coast from the growing threat of France and divert its forces from the Danube and Crimean directions. It was held during the war of the 3rd and 4th Coalitions against France. The squadron blocked the coasts of Adriatic sea occupied by the French, took part in the defense of the Ionian Islands, fought the French fleet on its communications, assisted the Adriatic people in their fight against Austrian rule and the French troops. After Turkey entered the war (December 1806), on the side of France, the Russian squadron, leaving a part of the forces in the Adriatic, moved into the Aegean Sea, where it conducted military operations against the Turkish fleet. On November 24, 1805, Alexander I, who faced the threat of Napoleon's invasion in Russia, ordered Russian troops to be returned from the Mediterranean theater of war, leaving small forces on Corfu. On December 14, 1805, a squadron was ordered to leave the Ionian Islands and pass to the Black Sea. However, D.N. Senyavin received this order only at the end of March 1806, when the naval and land forces led by the admiral launched active military operations in the Adriatic and achieved significant military and political successes. On his own initiative, D.N. Senyavin made a bold decision: taking advantage of the favorable political situation, capture the base on the east coast of the Adriatic Sea and prevent the transfer of Slavic areas to the French.

Adriatic Sea

Adriatic Sea – the semi-enclosed sea, which is the most extensive gulf of the Mediterranean Sea, with deep indraught into its northern coast between the Apennine and Balkan Peninsulas. It washes the shores of Italy, Slovenia, Croatia, Bosnia and Herzegovina, Montenegro, and Albania. The Otranto Strait, 75 km wide, is connected to the Ionic Sea. The International Hydrographic Organization (IHO) defines the southern boundary of Adriatic Sea as a line extending from the mouth of Butrinto river (Albania) to Karagol Cape on Kerkyra island (Greece) and through this island to Kephali Cape and to Santa Maria di Leuca Cape (Italy). Adriatic sea indraught to the shore at 770–796 km, the sea width is 93–248 km, and in the Bar area, 355 km. The total area is 138,600 km^2, which is 5.6% of the Mediterranean Sea total area. The large bays are Venice, Trieste, Manfredonia, and Boko-Kotor. There are more than 1300 islands in the Adriatic, mostly along the east coast. About 70 of them are uninhabited. They include 79 major islands, 525 islets, and 642 rocks. The islands located along the western Italian Adriatic coast are small and numerically insignificant comparing to those, which are situated at the eastern coast. Those 117 islands on which Venice is built are well known (Fig. 4).

Adriatic sea is shallow. The continental shelf (the outer edge of the continental shelf is located at a depth of 200 m) covers an area of 102,000 km^2, or 74% of the total area of the sea. Adriatic Sea gradually deepens from the northwest to the southeast. In general, the sea depth varies from 20 to 50 m in the northern part of the sea, up to 1230 m – in the south-east, with an average depth of 252 m. The total volume of Adriatic Sea is about 35,000 km^3. Adriatic western shores are predominantly lowlands, while the eastern shores are mountainous. Near the eastern shores the Dalmatian islands are situated. The

Fig. 4 Adriatic Sea. (Source: https://ontheworldmap.com/oceans-and-seas/adriatic-sea/large-detailed-map-of-adriatic-sea-with-cities-and-towns.jpg)

shoreline is jagged and rich with the best natural harbors. The shores of Adriatic Sea from the side of the Balkan Peninsula are the series of valleys flooded by the sea parallel to the coast – a Dalmatian type of coast, which explains an abundance of large and small islands.

Whereas the Italian coast of Adriatic Sea is smooth. Transparency of water in the sea reaches 40–50 m, therefore, it is one of the most transparent seas of the world. The main rivers discharging into the Adriatic sea are Po, Soča, Krka, Neretva, Drin, Bojana, and Vjosë. The catchment area of Adriatic sea is 235,000 km^2. The climate has Mediterranean specificities, however, differs significantly from the climate of the Mediterranean Sea itself. The area is characterized by the specific local winds (bora, mistral, sirocco), affecting air temperature. The average wind speed throughout the year in the most of the areas is about 3 m/s. The hydrological regime of Adriatic Sea is characterized by high water temperature, high salinity, slight coastal water dynamics, weak currents, and rare storms. The water temperature changes from north to south and varies in summer from 24 °C to 26 °C, in August and from 7 °C to 13 °C in February. Sea water salinity is approximately 38.3‰. There are constant tidal and currents observed at Adriatic Sea. Constant currents in Adriatic sea represent a cycle in a counterclockwise direction, which, apparently, is a continuation of the general circulation of water in Ionian Sea. The inflow of Ionic Sea waters into Adriatic Sea through the eastern part of the Strait of Otranto, determine the coastal north-western current. The special configuration Adriatic Sea is the reason explaining why the storms are more frequent, less expected, and are less predictable. The flora and fauna of the Adriatic Sea is quite rich – more than 7000 species. There are 407 species and subspecies of fish dwelling in Adriatic Sea. There are 353 species and subspecies related to Osteichthyes group and 54 species related to Selachii or Chondrichthyes group, which makes up 70% of the known species and subspecies of the fish dwelling the Mediterranean Sea (about 579 species and subspecies). There are more than 750 species of algae related to the three divisions: red, brown, and green. In the coastal zone there are many species of gastropods and bivalved mollusks with thick, solid shells, securing them from the impacts of waves, as well as echinoderms and crustaceans. Oysters, mussels, limpets, sea urchins, sea cucumbers, and small crabs dwell in shallow water. Within the thickets of algae swim the seahorses. Large crustaceans – lobsters, large crabs, as well as octopus, cuttlefish, starfish, dwell in deeper waters. Eels and moray eels also swim here. The water mass is abundant with plankton and young fish. The Sea is dwelt by flocks of sardines, bonetta, air bladder mackerel and frigate mackerel, bonito, and tunas. Currents bring many tender, transparent jellyfish, and hydroids gleaming in the night. The most common among the sharks are pygmy shark, spiny dogfish, blue shark, and also thresher shark. In rare cases, there may be met basking shark. Among the mammals, dolphins, and monk seals (q.v.) are endangered. In recent years, crawfish invaders (*Callinectes sapidus* and *Percnon gibbesi*) have appeared in the sea. Oil and gas are being mined on the shelf nearby the Italian coast. There is also conducted the exploration drilling, which helped to discover oil deposits on the continental shelf near the port of Vaspo and off the coast of Croatia. On the shores of the sea live more than 3.5 million people. The largest cities are Bari, Venice, Trieste, Ravenna, and Rimini (Italy), Split, Rijeka, Zadar (Croatia), Durrës, Vlorë (Albania), Koper (Slovenia), Budva, Bar (Montenegro). There are 19 large ports situated on the sea shores: Rijeka, Split, Pula, Dubrovnik (Croatia); Bar, Kotor (Montenegro); Koper (Slovenia); Durrës, Vlorë (Albania); Trieste, Venice, Bari, Ancona, Brindisi (Italy). The largest cargo seaport of the Adriatic Sea is Trieste, the largest passenger port is Split. In addition to the largest ports, there are numerous small ports that are important for local transportation (Šibenik, Zadar, etc.). The shortest sea and air routes between the Apennine and Balkan Peninsulas pass through the Adriatic Sea. Fisheries (sardines, mackerel) and tourism play a significant role in the development of the economy of the Adriatic countries. In these countries there is also developed mariculture (oysters, mussels). The Adriatic Sea, especially in the northern part, is one of the richest in fish in the Mediterranean basin.

Adriatic-Ionian Initiative (*AII*)

Adriatic-Ionian Initiative (*AII*) – the political initiative launched by Rome at the Conference on the Foreign Affairs Ministerial Level held in Ancona, Italy, from May 19 to 20, 2000. Alongside Italy, it was attended by representatives of Greece, Slovenia, Croatia, Bosnia and Herzegovina, and Albania. It is noticeable that a month before the conference in Ancona the project was referred to as the Adriatic Initiative, which was reflected in the joint Italian-Croatian declaration in April 2000. Apparently, the transformation of the name was caused by the desire of Greece not to stand aside in this project. The purpose of the initiative is the arrangement and post-crisis development in the Balkans, which could make the

Balkans one of the main centers of attraction for the countries of South-Eastern Europe in the future. Its structure consists of the Adriatic between the countries of the two seas for international cooperation and security for Albania, Bosnia and Herzegovina, Italy, Croatia, Greece, Montenegro, Serbia, and Slovenia.

"Aegaeo"

"Aegaeo" – a research vessel of the Greek Center for Marine Research. It was built in 1985 at the Chalkis shipyard. The vessel launched the studies in the eastern Mediterranean, and then in its western areas. The length of the vessel is 61.51 m, its width is 9.60 m, its tonnage is 778 t, its speed is 12 knots, and its autonomous navigation is 20 days. In 1996–1997 the vessel was re-equipped and modernized. Floating laboratories were equipped with modern facilities. The vessel carries two underwater vehicles – Tethys and Rov Super Achilles.

Aegean Sea (Egeo Pelagos)

Aegean Sea (Egeo Pelagos) – a semi-enclosed sea with a large number of islands (about 2000) in the eastern Mediterranean, between the Balkan Peninsula, Asia Minor, and Crete Island. The Aegean Sea is one of the cradles of Antiquity, ancient Greek and later Byzantine civilizations. It washes the shores of modern Greece and Turkey. In the northeast, through the Dardanelles, it connects with the Sea of Marmara, then through the Bosphorus Strait with the Black Sea, in the south through several straits between the islands – with the Ionian Sea and the Mediterranean Sea (Fig. 5).

The area of the sea is about 179,000 km^2. The banks are mostly rocky. The sea is framed by numerous, however, as a rule, low mountain ranges with a semi-desert landscape. The Aegean Sea was formed as a result of subsidence of the land (Aegeidas) at the end of Pliocene and in Pleistocene, the remains of which are numerous islands (in this regard, the Aegean Sea was previously called the Archipelago Sea). The largest islands are Crete, Lesbos, Rhodes, Samos, Euboea (or Evia). Depths of 200–1000 m prevail (the greatest depth is 2529 m in the south). The currents in the western part are directed to the south, in the eastern – to the north. Their speed is up to 0.5–1 km/h. The water temperature on the surface in winter is 11–15 °C, in summer 22–25 °C. Due to global warming, the sea water temperature and its salinity are constantly increasing. At a depth of more than 350 m, the water temperature does not change throughout the year (12–13 °C). The Aegean Sea is characterized by a higher salinity – 37.0–40.0‰. The tides are semi-daily, their size is 30–60 cm. Since olden times, fishing, sponge harvesting, and octopus catching have been developed in the Aegean Sea. Fishing has recently been gradually declining due to its private (semi-handicraft) nature and the deterioration of the ecological situation in the sea, as well as the poverty of plankton necessary for fish nutrition. The Aegean Sea is a traditional shipping region with a significant number of cargo and passenger ports. Main ports: Piraeus, Thessaloniki (Greece), İzmir (Turkey). Various civilizations developed in the Aegean Sea basin: Ancient Greece, Ancient Rome, Byzantine Empire, Bulgarian Kingdom, Latin Empire, Venetian Republic, Ottoman Empire. The name of the sea originated in ancient times, presumably from the ancient Greek aiges (αἶγες) – "wave," but a rethinking of the more ancient name is also possible. It has also been suggested that the sea was named after the city of Aigeus (Aigai), located on Euboea. Ancient Greek mythology connects the name of the sea with the name of the Athenian king Aegeus, who threw himself from a cliff into the sea, deciding that his son Theseus died in Crete, killed by the Minotaur. In the Aegean Sea basin, to this day, there is a low-intensity conflict between Greece (which controls almost all the islands of the Aegean Sea) and Turkey, which is trying to occupy a dominant position in the region's trade. Due to the fact that the Aegean Sea is of great strategic importance as a route for oil tankers from

Fig. 5 Aegean Sea. (Source: https://ontheworldmap.com/oceans-and-seas/aegean-sea/map-of-aegean-sea-with-islands.jpg)

the Black Sea, there are frequent oil spills and waste water discharge (especially in the Athens region). The floating hospital "Britannic" sank in the Aegean Sea, which was the twin sister of the famous "Titanic." The southeastern part of the Aegean Sea is called the Icarian Sea, the southern part of the Aegean Sea, the southern part of the Sea of Crete, and the northern part of the Thracian Sea.

Aetolia (Etolia)

Aetolia (Etolia) – a large region of Central Greece, with its southern side facing the Gulf of Corinth. From the west it bordered on Acarnania (q.v.), across the Achelous River, and from the east – with Locris (q.v.), Doris and Elis. The inhabitants of Aetolia were engaged in hunting and cattle breeding (horse breeding). The Aetolian tribes proved oneselves during the reign of Alexander the Great. In 370 BC they founded the Aetolian League, to which the Locrians and Arcadians joined. It represented a political counterbalance to Macedonia in central Greece. In 200–197 BC Aetolians fought on the side of Rome against Macedonia, and in 192 BC entered into an alliance with Antiochus III against the Romans, who in 146 BC turned their allied area into a Roman province.

Aetolia-Acarnania

Aetolia-Acarnania – one of the nomoi (q.v.) Central Greece (q.v.). This is a picturesque area in the western part of mainland Greece, which includes two most beautiful geographical regions – Aetolia and Acarnania, located along the banks of the Achelous River, which forms the natural border between them. Officially, it is called the nomos of Aetolia and Acarnania, and administratively it is included in the district of Western Greece. The region borders in the north-west with Lefkada, in the north with the districts of Arta and Preveza, in the north-east with Evrytania, in the east with Phocis, in the south with the region of Achaia. It is the largest district in Greece, but also the most sparsely populated. Population is approx. 223,000 people (2001). Most of the region's borders face the Ionian Sea. On the southwestern coast of Aetolia-Acarnania there are many lagoons, the most famous of which are Messolongi and Aitoliko. There is the largest lake in Greece within the region – Trichonida. The highest peak of Panaitoliko reaches 1924 m. There are mountain slopes with virgin pine, chestnut, beech forests, and turbulent streams rushing along them, fertile plains, peaceful and untroubled lakes surrounded by beautiful landscapes, rivers abundantly irrigating green valleys, groves and fields, and, which is the most important, endless kilometers of magnificent sandy coasts of the Ionian Sea with picturesque bays and rocky shores. The climate in the region is characterized by a high difference between summer and winter temperatures. It can be very hot here in summer. The temperature often rises to +40 °C. The humidity is high and it rains frequently. Winter in lowland areas is mild and warm, the air warms up to +10 °C on average. Winter is cold in the mountains. There are frequent snowfalls. In the highlands, even summers are cool. The inhabitants of the region are mainly engaged in agriculture and animal husbandry, which is the most developed in its southern part. Aetolia-Acarnania is one of the main tobacco-cultivation regions of the country. Rice, legumes, cereals, and cotton are also cultivated here. Salt and marble are mined. On the southwestern coast of the region, there are many fish farms where mullet is grown, which is used for caviar production. The region has one of the highest positions in the country for pork production. Many mountain dwellers are engaged in traditional sheep breeding. In these areas, agriculture is practically no longer practiced. In the mountainous regions of Nafpaktos, cows have been raised in recent years. Fruits, citruses, and olives are grown in the coastal areas. The nature of Western Greece takes a special place in the ecological system of the country. There are 2 of 11 wetland reserves located on its territory, which are of international importance, protected by the Ramsar Convention: the Ambracian Gulf and the natural complex of Messolongi with its lakes and lagoons. During the migration period, more than 300 species of birds stay here, among which there are endangered and rare species. Both national parks are included in the Natura-2000 system of the most important European ecological regions. According to legend, Aetolia is named after Aetolus, who settled here after fleeing from Akarnania, western neighbor of Aetolia. From the north, the region was adjacent to Thessaly and Epirus, from the east with Doris, and from the south it overlooked the Gulf of Patras (in ancient times it was part of the Gulf of Corinth). It was a mountainous wilderness, where even lions dwelled. The first inhabitants of northern Aetolia are the Leleges, and the southern ones are the Curetes and Boeotian Hyantes. The first Hellenic tribes – the Elians (Hellenes) – came, according to legend, together with Aetolus six generations before the Trojan War has broken out. In times of Homer, they were already called Aetolians. Until the Roman era, when the Romans annexed Aetolia to the province of Achaia, it was a large republic. There were not many cities in Aetolia. The largest of them were Chalcis, Makyneia, Elea (today's Missolonghi), Thermon, Trichonion, and Oechalia. Because of the isolated position of their country the inhabitants of Aetolia lagged far behind the rest of the Hellenes in culture and education. They were distinguished by a rude, warlike character. They often had conflicts with neighboring tribes. During the Lamian War, the Aetolian League (a confederation of tribal communities and cities) was created here, which, after the death of Alexander the Great, begins to play a significant role in the history of Greece. Its state structure was largely democratic. The power and influence of the Aetolians gradually spread to neighboring Locris and Phocis. It annexed part of Thessaly, Arcadia and Acarnania, and the Ionian Islands. Since the end of third century, their influence has become predominant in the Delphic Amphictyony. The Union reached its highest prosperity in around 220. It disintegrated in 189, when the Aetolians were defeated by the Roman consul Fulvius Nobilior, fighting on the side of Antiochus the Great. Acarnania, according to Apollodorus, was founded by the sons of Alcmaeon – Acarnanus and Amphoterus. In ancient times, it was not in any way significant in the history of Greece. The inhabitants of Acarnania, despite the fact that the area had a wonderful outlet to the sea, unlike all other Hellenes, were not interested in navigation. They were mainly engaged in agriculture.

Agios Ioannis (Ag. Ioannis)

Agios Ioannis (Ag. Ioannis) – an island 33 m high, lies west of the Zakynthos Island coast, Greece, 750 m from the entrance to Vromi Bay. The island is covered by maquis (q.v.). Sea cliffs are scattered between the Agios Ioannis Island and the Zakynthos Island.

Agios Nikolaos (Ag. Nikolaos)

Agios Nikolaos (Ag. Nikolaos) – an island 72 m high, 900 m length, and 550 m wide, located in the southern part of the Kerkyra Strait, east of the Mutros Island (q.v.), Greece. The island is 900 m long and 550 m wide. The island is separated from a coast by a narrow strait 180–260 m wide. This is very common name for small islands in Greece. Only in the Ionian Sea there are several such islands: (1) the island, located 3.5 km northeast of the northeastern coast of the Zakynthos Island (Katasgori Cape); (2) the rocky islet about 4 m high, located 200 m northwest of the caoe of the same name, bordering Argostolion Bay, the Kefalonia Island; (3) the island 15 m high, located 200 m southwest of the Pogonia Cape, Amvrakikos Bay. There is a chapel on the top of the island; the islet 56 m height, located in Gaios Bay, on the northeastern coast of the Paxi Island (q.v.). On the top of the island, there is the fort, the church Agios-Nikolaos is at the southern point of the island.

Agreement Between the Hellenic Republic and the Italian Republic on the Delimitation of the Respective Continental Shelf Areas of the Two States, Athens, 1977

Agreement Between the Hellenic Republic and the Italian Republic on the Delimitation of the Respective Continental Shelf Areas of the Two States, Athens, 1977 – Signed on 24 May 1977 in Athens, Greece. The agreement contains six articles. The division of the continental shelf zones is based on the principle of the median (middle) line (article I). The continental shelf boundary consists of 15 segments, 14 turning points, and 2 endpoints and runs from north to south for 260 nautical miles. The water depths at the border of separation vary from 800 to 4000 m. The northern end of the border is the point closest to the coasts of Greece and Italy – 22 and 20.1 nautical miles from the respective coasts. The southern tip is located 168.9 nautical miles from the coast of Sicily and 163.4 nautical miles from the Greek island of Stamfani. Article II deals with the problem of exploiting resources located on both sides of the boundary line.

Agreement on the Conservation of Cetaceans of the Black Sea, the Mediterranean and Contiguous Atlantic Area (ACCOBAMS)

Agreement on the Conservation of Cetaceans of the Black Sea, the Mediterranean and Contiguous Atlantic Area (ACCOBAMS) – "A collective instrument for the conservation of marine biological diversity in the Mediterranean and Black

Seas." ACCOBAMS aims to reduce threats to cetaceans in the Mediterranean and Black Seas, as well as to improve public awareness of these mammals. This is the first Agreement for the protection of cetaceans within the two regions, allowing them to cooperate on this issue. It was concluded within the framework of the Convention on Migratory Species in 1996 and entered into force in 2001. Precisely at this time, Greece made a decision for this agreement to enter into force. The company for the scientific research of spermwhales in the Ionian Sea, in particular, allowed to expand knowledge about the population of this species in the region, as well as to develop a methodology for assessing its total abundance in the Mediterranean basin, which is one of the priorities of the Parties to the ACCOBAMS Agreement. The permanent secretariat of ACCOBAMS is located in the Principality of Monaco.

Agri (Agri)

Agri (Agri) – a river in the southern Italy. The river flows in Basilicata region. The length of the river is 136 km, and the basin area is 1770 km^2. The river source is in the Lucan Apennines. It enters the Gulf of Taranto, the Ionian Sea, near the town of Policoro.

Agria Gramvousa (*Agria gramvousa*) ("*wild Gramvousa*")

Agria Gramvousa (*Agria gramvousa*) ("*wild Gramvousa*") – is one of two Gramvousa Islands (q.v.) is situated north of Imeri Gramvousa ("tame Gramvousa") (q.v.) at the northwestern end of the Crete Island, Ionian Sea, Greece. The island is uninhabited.

Agrostoli, Agrostalion (*Agrostoli*)

Agrostoli, Agrostalion (*Agrostoli*) – (1) a town, the capital of the Kefalonia Island, Greece. It is located at the eastern entrance into the bay-gulf of Agrostoli. The locals call it "Piccolo Paris." Population 9748 people (2011). Until 1953, Agrostoli was a small fishing village (Fig. 6).

In 1953, the town was almost devastated by the catastrophic earthquake and rebuilt, but the most part of architectural monuments were not restored. In 1943, after Italy came out of the war on the side of Hitler, nine thousands of the Italians from the Alpine Division refused to surrender the island to the Germans and continued to resist during seven days. After that, 3000 former

Fig. 6 Agrostoli. (Source: https://www.hotels4all.gr/wp-content/uploads/2016/05/ARGOSTOLI-IMG_3858.jpg)

allies were captured and executed by the Germans. The memorial and monument, which were erected near Agrostoli, commemorate the Italians executed by the Nazis. The stone multiple-arch bridge, connecting two coasts of the internal gulf (limnophase), was constructed by the British in 1813. The locals name it "Drepano" (bow, sickle). Here, there is the Maritime Academy. In terms of sightseeings, there are the archeological museum, historical-ethnographical museum Korelianeo, and St. Theodores lighthouse (q.v.). In the town, there is an airport and continuous traffic with the continent and the Ithaka and Zakynthos Islands. (2) a bay, which enters into the southern coast of the Kefalonia Island, the Ionian Sea, Greece. The northern part of the bay is called Livadi. Agrostoli Bay is narrow, but with a wide entrance. The eastern coast of the bay is higher and steeper than the western one. The mountain ridges stretch along the eastern coast. The southern part of the western coast is relatively low, but gradually rises to the northwest. The Vardiani Island is located at the entrance of the bay. The Agrostoli port is arranged at the eastern coast of the bay, the Lixsourion harbor (Liksouri) at the western coast.

Aitoliko (Etolikon)

Aitoliko (Etolikon) – (1) a lake-lagoon, in the northern part of Missolonghi Lagoon (q.v.), near the coast of Gulf of Patras (q.v.). It is connected to the lake by a narrow strait, the length of which is 3.5 km. The lake is 7.5 km long and 3.7 km wide. The shores of the lake are high. A railway runs along the eastern coast to the town of Messolonghi. From the port of Messolonghi to the city of Aitoliko, a canal with the smallest depth of 0.3 m runs along Missolonghi Lagoon; (2) an islet located in the middle of the entrance to the lake, on which the city of Aitoliko, founded in the Middle Ages, is located. There are bridges across the island from the east and west. One of them is 300 m high, made of stone, arched. It is famous for its rich maritime traditions.

Albania, Republic of Albania (*Shqipëria, Republika e Shqipërisë*)

A

Albania, Republic of Albania (*Shqipëria, Republika e Shqipërisë*) – is a state located in the western part of Balkan Peninsula in southeastern Europe. The name of the country comes from the Illyrian olba – "village." In the Middle Ages, the country was called "Arber," and from the sixteenth century, it was called Shqipëria. The self-name Shqipëria means "Country or Land of the Eagles" (an eagle in Albanian is shqiponjë). According to another version, this name comes from the Albanian word shqip – "speak clearly." The capital of the country is Tirana (557,400 people), the other largest city is Durrës (Figs. 7 and 8). The Strait of Otranto separates Albania from Italy. Albania shares its borders with the partially recognized Republic of Kosovo to the northeast (112 km), with Montenegro to the north-west (172 km), with Macedonia to the east (151 km), and with Greece to the south (282 km). On the west it is washed by the Adriatic Sea, and on the extreme south-west – by the Ionian Sea. Albania is stretched over 340 km from north to south and 150 km from west to east. The area of Albania is 28,748 km^2. The coastal length is 427 km, 273 km of which is washed by the Adriatic Sea, and 154 km by the Ionian Sea. The western part of the coast, stretched along the Adriatic Sea, is presented by the lowlands, and the rest area (about 70%) is presented by mountainous areas with an average height of 700 m. The highest point is Mount Korab (2764 m) in the eastern part of the country. The most famous beaches of the Adriatic coast of Albania are Velipoje, Durres, Hawaiian beach on the coast of Kavajë, Shëngjin, as well as Lalsit Bay. The climate of the coastal part of the country is Mediterranean, becoming continental in the eastern part. The average January temperature is 8–9 °C, and in July it is 24–25 °C. The amount of precipitation is 800–2000 mm per year. The total volume of renewable water resources is about 42 km^3. Rivers in Albania flow through deep valleys and have a mountainous specificity with a large hydropower potential. The rivers are mainly rain-fed. The

Fig. 7 Albania. (Source: https://www.maps-of-the-world.ru/maps/europe/albania/political-and-administrative-map-of-albania-with-roads-major-cities-and-airports-small.jpg)

Fig. 8 Albania. (Source: https://maps.spravka-region.ru/img/312/2438850776935314.jpg)

largest river of Albania is Drin (280 km) which flows through the northern regions of the country. Among other rivers are Mat, Vjosë, Shkumbin, and Seman. Almost all of the rivers flow into the Adriatic Sea. Three largest and deepest tectonic lakes of the Balkan Peninsula are partially located in Albania. Lake Shkodër (Lake Skadar), the third part and 57 km of the coast of which belong to Albania, is located in the north of the country, on the border with Montenegro. Lake Ohrid is divided between Albania and the Republic of Macedonia. The lake is characterized by a depth of 289 m and unique flora and fauna, and it is protected by Albania, Macedonia, and Greece, and lies somewhat further south. There are more than 3000 of vegetation species. Maquis is widely spread on the coast. About a third of the territory of Albania (mostly mountainous areas) is covered with forest, which is inhabited by brown bears, wolves, jackals, lynxes, and forest cats, as well as wild boars, roe, and deer. In coastal areas, some of which are lagoons, there are many water birds. There are deposits of oil, natural gas, coal, chromium, phosphates, copper, nickel, and salt in Albania. Albania is a parliamentary republic. The head of state is the President. The parliament of the country is the one-chamber National Assembly (140 seats). The executive authority is the Council of Ministers. The Kingdom of Albania was known since February 1272. Albania was kept under the control of the Ottoman Empire for five centuries. The country gained its independence from it in 1912. During the First World War, Albania was occupied by Serbia and Greece. Under the Second Treaty of London in 1915 with Italy, the Allies made a secret promise to turn Albania into a protectorate of Italy. The first parliament of Albania was created in 1920 during the struggle for the independence of the country and against its partition according to Paris Peace Treaty between Greece, Italy, and Yugoslavia. In 1928, the parliament was dissolved, Albania was declared a monarchy. In April 1939, fascist Italy occupied Albania. In September 1943, it was occupied by Hitler's Germany. As a result of the national liberation war of the Albanian people (1939–1944), in November 1944, during the Soviet Army offensive in the Balkans, the country was liberated from the occupiers. In 1945, parliamentary elections were held, in which the Democratic Front headed by the Communists won 97.7% of the vote. In 1946 Albania was declared a national republic. Gradually, Enver Hoxha – Albanian Stalin, concentrated power in his hands. Until 1956, Albania maintained relations with the USSR. However, after the 20th

Congress of the Communist Party of the Soviet Union (February 14–25, 1956), a policy of political isolation was adopted. Relations were maintained only with China and Romania. In December 1976, the People's Socialist Republic of Albania was declared. In the early 1990s Albania began a policy of cautious economic reforms and expansion of relations with other countries. In 1990, a multiparty system was adopted. In 1991, the name of the state was changed to the Republic of Albania, the democratization of the country and the transition to a market economy began. Since April 2009, Albania is a NATO member, a member of the UN, OSCE, COE, and since 2009 an associate member of the European Union. The territory of Albania is administratively divided into 12 prefectures or counties ("Karks"), but, as a rule, Albania is divided into two parts – the northern (Ghegeria) and the southern (Toskeria). Six prefectures of Albania overlook the Adriatic Sea coast: Shkodër, Lezhë, Durrës, Tirana, Fier, and Vlorë. The population is 2.9 million people (2017). Ethnic composition is 95% Albanians, 3% Greeks, and others (Wallachian, Gypsies, Serbs, Macedonians) 2%. Among the believers there are about 60% of Muslim. About 53.5% of the population lives in urban areas, 46.5% in rural areas. Albania is an agrarian-industrial country. Economic structure is the following: agriculture contains 20.6%; industry contains 18.8%; service sector contains 60.6%. In recent years, tourism is being developed. The industry includes the following areas: food processing, textiles, and clothing; wood processing, oil extraction, chromite mining, copper mining, iron-nickel ore mining, brown coal mining, natural bitumen mining. There is also concrete production and hydropower. Wheat, corn, potatoes, vegetables, fruits, sugar beets, and grapes are grown in Albania. Animal husbandry is also developed. Goats, sheep, and cattle breeding are developed. There is also tobacco production. Albania exports: chromites, crude oil; asphalt, vegetables, tobacco products. The main buyers are Italy, Greece, and China. Albania imports: machinery and equipment, food, textiles, and chemical products. The main suppliers are Italy, Greece, Turkey, Germany, China, and Russia. There are four sea ports: Durrës, Vlorë, Shengjin, Sarandë. There is a naval base in Vlorë. There is one international airport in the country – "Mother Teresa," located in 20 km away from Tirana in Rinas. The road transport is developed. The railway network is represented by two roads – Tirana-Durrës (built by Soviet engineers) and Durrës-Pogradec. It is interesting that there are about 600,000 of permanent fire positions in the country – bunkers built in almost every yard during the dictatorship of E. Hoxha.

Albanian Navy (*Forcat Detare Shqiptare*)

Albanian Navy (*Forcat Detare Shqiptare*) – one of the branches of the armed forces of the Republic of Albania. The Navy (headquarters in the city of Tirana) consists of a Fleet and coastal defense forces, geographically divided into two regional naval areas (RNR). The first RNR (headquarters in the city of Durrës) includes the Durrës naval base and the Schengin base station (BS). The second RNR (headquarters in the city of Vlorë) includes the naval base Vlorë and BS Saranda. The composition of each RVRM includes the division of two divisions of combat boats. The combat structure of the Navy includes four warships, 20 combat boats (torpedo, patrol), two divisions of minesweepers, and two auxiliary vessels. The total number of troops is 2000 people (Fig. 9).

Alfeiós (*Alphios*)

Alfeiós (*Alphios*) – is the longest river on the Peloponnese peninsula, Greece. Its length is 120 km. The catchment area is 3600 km^2. It originates on the slopes of the Taygetus mountains and flows into the Gulf of Kyparissia in the Ionian Sea. The water flow is uneven with heavy floods during the winter rains. The river is unnavigable. However, ships with a draft of up to 1.2 m may sail up the river at 5.5–7.5 km.

Fig. 9 Albanian Navy. (*Forcat Detare Shqiptare*) (Source: https://dic.academic.ru/pictures/wiki/files/80/Project122bis-2007-Pashaliman.jpg)

Alkantara

Alkantara – is a river in the east of Sicily, Italy. The length of the river is 53 km, the basin area is 573 km^2. The water flow is 9 m^3/s. The river has its source on the southern slope of the Monti Nebrodi at an altitude of about 1393 m. The river flows eastward, and has its mouth in the Ionian Sea. The competence of the river varies greatly depending on the season. It dries up often.

Alkyonides (*Alkionides*)

Alkyonides (*Alkionides*) – islands located in the middle part of the Alkyonides Gulf (q.v.), Greece, occupying its part, which is about 3.6 km long and 2.4 km wide. The islands consist of four small high islands: Zoodochos Pigi, Daskalio, Glaronisi, and Praso. The passages between them are narrow and mostly dangerous. The width of the passage between the islands and the north shore of the bay is 4.6 km, with a minimum depth of 350 m. The width between the islands and the southern shores of the gulf is characterized by the width of 3.1 km and the lowest depth, which is 440 m.

Alkyonides Gulf (*Alkionidon kolpos*)

Alkyonides Gulf (*Alkionidon kolpos*) – is a gulf which is the eastern part of the Corinthian Gulf (q.v.), and extending into the coast at 28 km, Greece. The northern coast is heavily indented. At a distance of less than 1.8 km there are separate mountains up to 970 m high. Two bays, separated by a mountainous peninsula, extend into the east coast. The southern shore of the bay is low and considerably indented. It is the northern coast of the Corinthian isthmus (q.v.), connecting the Peloponnese peninsula with the mainland. In the middle of the bay there is a group of Alkyonides (q.v.). The depth at the entrance to the bay and in its middle part is 300–570 m.

Ambracia (*Ambrakia*)

Ambracia (*Ambrakia*) – a Corinthian colony with a heavily fortified port in the north of Ambracian Gulf in the Ionian Sea on the bank of the Arachthos (q.v.), Greece. In 426 BC it was conquered by the acarnanians allied with Athens.

Pyrrhus from Epirus made Ambracia his residence. Later, Ambracia became a free city.

Ambracian Gulf (Amvrakikos kolpos)

Ambracian Gulf (Amvrakikos kolpos) – is a gulf, which extends into the coast of the Balkan Peninsula in 9 km north from the northern entrance to the Lefkas canal. The gulf is connected with the Ionian Sea by wandering shallow Preveza Strait (q.v.) in the lowland strip, which is 2.7–3.7 km wide. The area of the gulf is 654 km^2. Its length is 40 km, and its width is 15 km. The southern coast of the bay is predominantly high, and only its western part is low and marshy at the entrance to the gulf. Three large bays extend into the coast. In some places, the shore is bordered by shoals with 370–550 m wide. The North Coast is heavily indented, low and mostly swampy, there are a shallow lake and lagoons. The gulf is the mouth for Arachtos River (q.v.). Two bays extend into the coast. The ancient city of Nicopolis is situated on the coast. The shore is also bordered by shoals with the width up to 2 km in some places. The bottom topography is rugged.

The depth in the middle part of the gulf varies from 22 to 60 m. Only fishing vessels and vessels supporting communication between settlements on the shores ply in the bay. It is the most important protected wetland area in Greece. There are about 300 species of birds found, 71 of which are the species listed in the Red Book. It is known for a mullet caviar or ambracian caviar. Also it is famous for the naval battle near the promontory of Actium (31 BC) (q.v.). Such city as Arta (at the mouth of the Arachtos River), Koronisia village, Menidi village, towns of Amfilochia and Vonitsa are located.

Anapo

Anapo – is a river in the southeast of Sicily, Italy. The length of the river is 40 km. The headwaters of the river is located in the Hyblaean mountains, not far from the top of Monte Lauro. It flows to the East through the Siracusa province and flows into the Ionian Sea near the city of Siracusa. In many places the waters of the river run deep into the earth, which is why the river got its name. In Greek language its name means "invisible."

Anchovies (*Eengraulis*)

Anchovies (*Eengraulis*) – a genus of marine fish from the family of Engraulidae. Anchovies dwell in some distance from the coast and will never enter the open ocean. They also may be met in waters with temperature varying from +6 °C to +22 °C. Anchovies make daily (vertical) and seasonal migrations. The maximum length of Anchovies does not exceed 20 cm. However, the abundance of these schooling fish is very large (Fig. 10).

According to the total abundance of all species, they take the first place among the fish. Anchovies are plankton-eaters, playing an important role in marine trophic chains. In turn, they are a significant food source for predatory fish, dolphins, squids, and seabirds. Their population fecundity is 20,000–30,000 eggs. Anchovies are one of the most important groups of commercial fish. The value of anchovies as a fishing object is largely determined by the high fat content of these fish. They are characterized by the high eating qualities. During the Antiquity, Mediterranean

Fig. 10 Anchovies. (Source: http://pitall.ru/wp-content/uploads/2017/09/anchousy.jpg)

Anchovies were highly valued in pickled form and for cooking Garum sauce (which is "Garos" in Greek).

Antikira, Andikira kolpos

Antikira, Andikira kolpos – is a bay, which extends 15 km to the northern coast of the Corinthian Gulf in the Ionian Sea, Greece. The coast is formed by the southern and southeastern spurs of Mount Parnassus, which is 15 km away of the top of the bay. The shores of the bay are high and heavily indented. Three relatively large bays and several small bays extend into the coast. The western shore, which is about 13 km long, is characterized by a general direction to the north-north-east. In the southeast of the western shore, there lies the Kephali peninsula, which separates the top of Antikira from its middle part. The eastern coastline of Antikira with a length of 18.5 km is higher and less indented than the western one. The depth of the bay at the entrance and in the middle part exceeds 150 m.

Antikithira

Antikithira – a tiny island located 13 km south of Kythira island. Part of the Ionian Islands, Greece. It is rocky island. Until 1964, political prisoners were exiled here. Population of the island is 45 people. The large settlement is Potamos, which is a port at the harbor. In the harbor there is a well-preserved antique slipways in the form of a trench carved in stone, through which ships were launched. In 1902, a Greek archaeologist Valerios Stais discovered a shipwrecked antique ship near this island, and among its fragments there was a bronze mechanism was created in about 100–150 BC (according to other sources – 87 BC), which was called "The Mechanism from Antikythera" (Fig. 11).

Some believe that this device automatically calculated the movement and position of the Earth, the Moon, and five other planets. Other people, from the scientific community, beleived that this is the first analog computer in the world. The artifact is kept in the National Archaeological Museum in Athens.

Fig. 11 "The Mechanism from Antikythera". (Source: https://pentad.ru/wp-content/uploads/2019/09/Antikiterskij-mehanizm.jpg)

Antipaxos (Andipaxios)

Antipaxos (Andipaxios) – is an island, one of the smallest isles in the Ionian Sea in Greece. It is situated south of Kerkyra (Corfu) and in a close neighborhood (2.2 km) to the southern part of Paxos island (q.v.). Area of the island is 5 km^2. Population of the island consist of 64 people (2001). The surface of the island gradually rises to the north and reaches an elevation of 112 m. The coastline is mostly indented and cliffed. There is only one settlement on the island. Antipaxos is largely covered by vineyards. There are two beaches here: sandy Vrika and pebbled Voutoumi and harbor "Agrapidia." On the south-eastern tip of the island there is a lighthouse. In about 1 km from a lighthouse, there is Dascalio island, surrounded by underwater cliffs. The island is separated by a strait from Paxos island, where the lowest depth is 20 m (Fig. 12).

Fig. 12 Antipaxos Island. (Source: https://i.ytimg.com/vi/mbaOnZsYMCA/maxresdefault.jpg)

Apennine Peninsula, Italian Peninsula, (Penisola Appenninica)

Apennine Peninsula, Italian Peninsula, (Penisola Appenninica) – is the medium one among the three large peninsulas of Southern Europe (the others are the Balkan and Iberian), occupies a central position (extends into the Mediterranean Sea), dominating along with the Island of Sicily over the communication routes between the Eastern and Western Mediterranean. Received its name from the Apennine Mountains, has the nickname "Boot." The three smaller peninsulas that make this distinctive appearance are called Calabria, Salento, and Gargano. The Apennine Peninsula is washed by the waters of the Mediterranean: from the west by the Ligurian and Tyrrhenian seas, from the east by the Adriatic Sea and from the south by the Ionian Sea. In the south, it is separated from Sicily by the narrow Strait of Messina (3.5–22 km). The area of the Apennine Peninsula is about 149,000 km^2. Its length is approximately 1100 km, and width varies from 130 to 300 km (Fig. 18). In the north it is bounded by the vast fertile Padan Lowland of the Po River, swept from the north by the powerful arc of the Alps. Along the central part of the Apennine Peninsula the mountain ridge of the Apennines (the maximum height of the Mount Corno is 2912 m) passes. Apennine mountains do not occupy the entire peninsula. The north of the Apennines stretches along the coast of the Gulf of Genoa, bordering the south of Padan Plain. The narrow strip between the mountains and the sea is called the Riviera: the French – in the west, and the Italian – in the east. Within the peninsula the Apennines deviate to the south-east and stand quite far from the Tyrrhenian Sea. To the upper reaches of the Arno River, the mountains are called the Northern Apennines. In this part they are composed of Paleogene, mostly loose rocks and rarely exceed 2000 m. The predominance of clayey deposits in the structure of the Northern Apennines creates conditions for the development of landslide phenomena, which are intensified due to the destruction of forests. Many residential places in the Northern Apennines are located in deep tectonic depressions. In one of these basins lies the ancient city of Florence. To the south, the Central Apennines are composed of Mesozoic limestones and fall into high massifs separated by deep depressions and tectonic valleys. The slopes of the massifs are mostly steep and bare.

The highest parts of the mountains experienced glaciation, and their glacier forms are clearly expressed in their topography. Forest depletion contributed to a very strong development of karst formation in the Central Apennines. In the very south of the Apennines extremely close to the Tyrrhenian coast and in some places plunge directly to the sea. The activity of the tidal bores developed in the limestones specific topography forms. In the west and east between the mountains and the coast lies a strip of flat or hilly topography; the structure is not related to the Apennines. On the west coast, steep shores alternate with low ones. There are also many bays and gulfs. The east coast in the north and in the central part is a lowland; it is marshy, with abundance of lagoons. The coastal stripes of Tyrrhenian and the Adriatic Sea on the Apennine Peninsula are characterized by the different structure and topography. The strip along the coast of the Tyrrhenian Sea reaches its greatest width in the north, where among the low, hilly plains rise the separate crystalline massifs – part of the same ancient land as the mountains of Calabria. Further to the south, the ancient and young volcanic formations begin to play a significant role in the structure and topography of the area adjacent to Apennines. There are a number lost volcanoes rising and plains stretching composed of volcanic rocks and dissected by rivers. On the hilly volcanic plain stands the capital of Italy, Rome. In many places of this area there are hot spring outlets. Further south, in the area of Naples, a double cone of Vesuvius, one of the most active volcanoes in Europe, rises. The vast areas around Vesuvius are covered with lava, which has poured out during numerous eruptions, and is covered with masses of volcanic ash. The most severe volcano eruption at the beginning of our era led to the death of three cities: Pompeii, Herculaneum, and Stabiae (79 AD), located at its foot. As a result of the excavations that began in the last century, Pompeii was partially liberated from under the strata of volcanic rocks that covered them and turned into a museum-reserve, widely known throughout the world. Stromboli Volcano on Stromboli Island located north of Sicily is active continuously throughout the entire historical time and was called the "lighthouse of the Tyrrhenian Sea," the highest and most active volcano in Europe, with a height of 924 m. Strong earthquakes are not uncommon for the Apennine Peninsula. From the side of the Adriatic Sea at the foot of the Apennines lies a raised hilly strip, which is called Subapennin. In the southern part of Subapennines, they turn into a karst limestone plateau up to 1000 m high, which stretches from the Gargano Peninsula to the Salento Peninsula. On the Apennine Peninsula the main part of Italy, the Republic of San Marino and the Vatican state are located. The climate of the peninsula is Mediterranean, especially in the western part. However, high altitude climate zone is of great importance, so that the actual Mediterranean climate is specific only for a relatively narrow coastal strip. From north to south, there is also a significant change in climatic conditions. The most typical Mediterranean climate is specific for the islands, especially for Sicily, as well as for the southern part of the Apennine Peninsula. The orography has a great influence on the climate, especially on the distribution of precipitation. In general, in the west the climate is warmer and more humid than in the east. The climatic specificities of the region largely depend on the Alps rising to the north of it, which are an almost formidable obstacle to the invasion of cold air from the north. Only in rare cases, at intervals of several decades, when an unusually severe winter begins in Western Europe, the cold air masses pass through the Alps or flow around them and spread far to the south. At the same time, there are frosts and snow all over the Apennine Peninsula. Especially soft is the climate of the coast of the Ligurian Sea – Riviera. This narrow coastal strip lying close to the sea from the north is protected by mountains from the invasion of cold air masses. Winter is usually warmer here than in the more southern regions of the Apennine Peninsula (average January temperature +8 °C); precipitation is abundant – up to 3000 mm, and their maximum falls in autumn. Summer is sunny and dry, intense heat is moderated by the proximity of the sea. Frosts on the Riviera are very rare phenomena, as well as the snowfalls. In the northern part of the peninsula, the climate is not so much mild as on the Riviera. The average January temperature in Florence and Rome is +5 °C and +6 °C, with frosts and snowfall occurring every year. The amount of precipitation in the west exceeds 1000 mm, and in the east it is usually no more than 500 mm; their maximum occurs in the fall and spring, when the polar front passes through these areas. The average July temperature is +24 °C and +25 °C. The climate of Calibria is much warmer, and Sicily is one of the warmest places in Europe. The average January temperature in Sicily is +11 °C or +12 °C, precipitation does not exceed 600 mm, they fall out within three winter months, and all other seasons are rainless. The climate gets cooler as we move up the mountains. In the Apennines at the latitude of Rome, the snow lies from November to March, and the slopes of Etna above 2000 m are covered with snow for a significant part of the year. The region is not rich with surface waters, and therefore, there are no long navigable rivers. The largest river is the Tiber (400 km long), on which stands the capital of Italy, Rome. The river regime is uneven and very clearly reflects the climate. All streams become shallow in the dry summer period, sometimes completely dry. They overflow with water during the winter and in the north during the autumn–spring rains. During especially heavy rainfall falling simultaneously in different parts of the basin, catastrophic floods occur. Thus, during the severest flood in November 1966, the Arno River rose several meters and flooded Florence with a layer of liquid mud. As a result, many art masterpieces, which were stored in the museums of this city, were lost. The Apennine Peninsula and the adjacent islands are characterized by a vegetation diversity, especially at a small height above sea level. The elements specific for the temperate zone still predominate in the natural and agricultural vegetation cover in the north. Only in the southern half of the peninsula and on the islands the vegetation and soils acquire a typically Mediterranean character. Some exception is the Riviera. It is one of the largest resort areas in Europe with rich decorative vegetation. The alleys of palm trees, groves of orange, olive and fig trees, vineyards, planted groves of cork oak, roses, and other flowering aromatic plants stretch along the coastline. On the shores of the northern part of the island, the vegetation and soils do not have manifested Mediterranean specificities. Deciduous trees and shrubs play an important role in the composition of the flora. Typical maquis and evergreen forests appear in the west around the area of Livorno, and as for the east, they occur only on the south of Ancona. As a rule, shrub vegetation is common at the site of destroyed forests, and primary shrubs in many places have given way to diverse crop vegetation. The natural evergreen forests have almost completely disappeared. In many places, groves of artificially planted Italian stone pines or evergreen oaks are common. Tangles of oleanders, brightly blossoming in spring, stretch along the river valleys. Among the cultivated plants the olive tree, grains, and vines dominate. Citrus plantations begin at the south of Rome, with the Calabria Peninsula best known for their orange groves. In the mountainous areas at an altitude of 500 m in the north and 1000 m in the south grew deciduous and mixed forests on brown forest and podzolic soils. However, most of these forests are carved out. The bare slopes of mountains are observed from everywhere. In these places, the intensive processes of soil slump and erosion occur. All of these processes are the result of deforestation, which in some places becomes catastrophic. On the tops of the highest mountains there are areas of alpine meadows; however, these areas are relatively small. On the slopes of Etna, there is a specific alpine relic vegetation. There are widespread tangles of shrubs, which include birch – a relic of the ice age. Due to the intensive forests devastation, wild fauna has been relatively poorly preserved. There are almost no large mammals. In rare cases, in the Apennines, chamois may be met. There are also wild sheep mouflon and the Sardinian wild mountain goat. Among the small mammals there is a ferret, wild cat, hedgehog, and hare. There are a lot of different birds and

Fig. 13 Apennine (Italian) Peninsula. (Source: https://upload.wikimedia.org/wikipedia/commons/thumb/c/c6/Italy_Map.jpg/1200px-Italy_Map.jpg)

reptiles. Coastal waters are rich in fish. Tuna and sardines are especially appreciated. The region is not as rich in minerals as other parts of southern Europe. There are almost no reserves of mineral fuels; however, there are small deposits of iron ore. The latter are concentrated in the Tyrrhenian Alpine region. The crystalline rocks of many regions are a valuable building and cladding material (Fig. 13).

Apulia (Puglia)

Apulia (Puglia) – a region in southern Italy, the southeastern brow of the Apennine Peninsula with Cape Otranto, the easternmost point of the country. It is characterized by the longest coastline among continental areas, washed from the east by the Adriatic Sea and the Strait of Otranto and from the southwest by the Gulf of Taranto of the Ionian Sea. Apulia is an "Italian shoe heel." It is bordered by the areas of Basilicata, Campania, and Abruzzo to the west. It includes the provinces of Bari, Brindisi, Lecce, Taranto, and Foggia. Its area is of 19,360 km^2. The main cities are Bari, Taranto, Brindisi, Foggia, and Lecce. The population is 4.1 million people (2016) (Fig. 14).

Plains (53.3%) and hills (45.3%) are specific for Apulia, 1.5% of its territory is covered by mountains, which makes the region the least mountainous in Italy. Forest area contains only 5%. The predominant vegetation is an evergreen shrub (maquis). The Gargano massif (over 1000 m), the Tavoliere Plain, Le Murge Plateau, and the Salento Peninsula stand out due to their natural conditions. The limestone Gargano massif is mountainous in the inner part of its area and marshy near the coast. It is not suitable for settlement and economic activity. The flat Tavoliere Plain adjacent to Gargano massif from the west is characterized by its fertile soils, and thus, it is the breadbasket of Apulia. The hilly Le Murge Plateau is abundant with deep depressions and hollows, and it is riverless. It is situated to the south from Gargano massif. The Salento Peninsula is a hilly lowland, almost entirely used for vineyards, olive plantations, and orchards, including industrial crops. The climate is Mediterranean, and the annual precipitation is 500–600 mm. There is little surface water. The main river is Aufidus River (at the present it is Ofanto River). There are orchards and vineyards on the irrigated areas. Apulia is one of the main agricultural areas of Italy. About 30% of the cultivated area is occupied by orchards, olive, and fruit trees. Horticulture is developed on the coast, and in some places truck gardening. More than 50% of the population are employed in agriculture. Apulia takes the first place in Italy for the grapes harvest and the wine production, and for the olives harvesting and olive oil production. It takes the second place for tobacco and almonds harvesting. There is significant cereal production (wheat, oats, barley) in Apulia, which was a traditional baker in ancient Rome. Horticulture and melon-growing (mainly on the Adriatic coast) are developed. As well as in

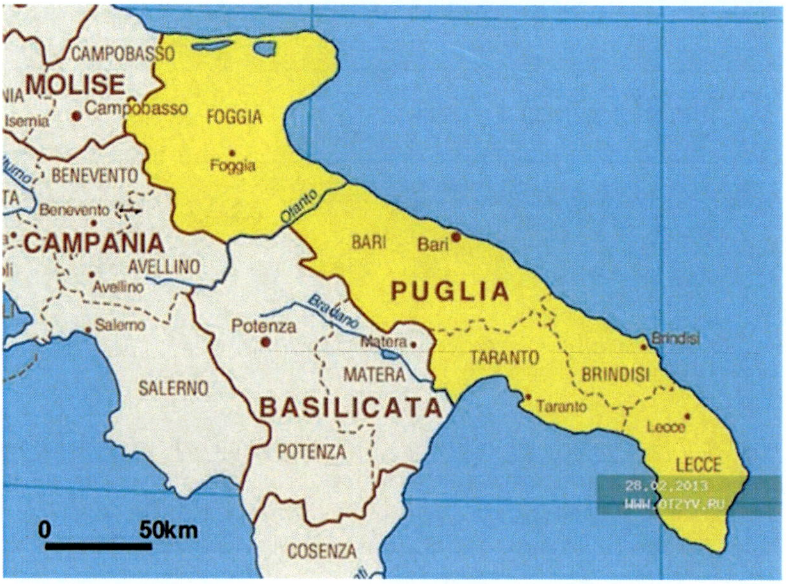

Fig. 14 Apulia Region, Italy. (Source: https://f.otzyv.ru/f/10/10/63553/14622/2802131746030.jpg)

the most of southern regions, the industry there is poorly developed. In Apulia there are the largest salt mines in Italy (Margherita di Savoia) and bauxite mines (Gargano Peninsula). Among the manufacturing industry, the most important are wine making, olive oil production, tobacco, and flour-grinding industry. There are separate enterprises of machine building (shipbuilding), woodworking industry, metallurgy (Taranto), petroleum refining (Bari, Brindisi), petrochemical industry, fertilizer manufacturing, papermaking, garment industry, textile industry, and rubber industry (tire works in Bari). The main industrial centers and ports are Bari, Taranto, and Brindisi. In Taranto, the second most important naval port of Italy, there are shipyards and an arsenal. The ancient dwellers of Apulia were aveons (or apuls) who originally belonged to the local tribes, and later to the Illyrian ones. During the times of Augustus Apulia belonged to the second region of Italy. Diocletian transformed Apulia into the province. Appia and Trajan roads passed through Apulia. In the eighth to fifth centuries BC, Apulia was under the Greek colonies. It was the Greeks who founded the cities of Luceria, Venusia, Canusium, Sypontum, Gardenia, etc., as well as the ports of Brundisius, Bari, and Hydrup. The land was severely ravaged by the Romans during the times of the Samnite and Punic Wars. At the end of the Republic and during the Age of Empire, Apulia was a district of large extensive livestock farms, where large masses of slaves were concentrated. In 568, Apulia was conquered by the Lombards. In the first half of the ninth century the Arabs were established there. Later, however, they were pushed out by the Byzantines. A large role in the struggle against the Normans who raided Apulia was played by the Russian (so-called Varyag-Russians) troops (druzhinas), who were at the service of Byzantium. In the eleventh century, the Normans, led by Guilliame the Iron Hand, occupied Apulia. The Normans chief, Robert Guiscard, who received the title of Duke of Apulia in Calabria from the Pope, in 1071, captured the Byzantine stronghold in Italy – the city of Bari. In 1127, under Roger II reign, Apulia became a part of the Kingdom of Sicily. In the thirteenth century together with Southern Italy Apulia was passed to Charles of Anjou and became a part of the Kingdom of Naples, which in 1442 was conquered by Aragon after the unification of Aragon and Castile into a single Kingdom of Spain. Since 1502, Apulia has been a part of the Spanish Realm until 1713, when the Kingdom of Naples was separated from Spain. As a member of Kingdom of Naples, Apulia joined Italy. In 1861 the region became part of the Kingdom of Italy, with the new capital city in Turin.

Arachthos

Arachthos – is a river entering the Ambracian Gulf of the Ionian Sea, Greece. The river is quite wide, with a meandering stream bed. It flows through the bogged lowland. It carries clay, which is carried and shifted to the mouth of the river. Small vessels are able to navigate the river, and sometimes sail up to the town of Arta, located on the left bank 20 km away.

Archimedes (*about 287–212 BC*)

Archimedes (*about 287–212 BC*) – is an outstanding antique mathematician and physicist. After studying in Alexandria, returned to Syracuse (q.v.). Determined the area (quadrature) of the parabola for the purposes of the geometric series summing, determined the name of integers. He used such mathematical methods, which are partially equivalent to the modern integral calculus. As a physicist, he justified the law of leverage and discovered the basic law of hydrostatics (Archimedes' law) (Fig. 15).

During the defense of the city of Syracuse during the Second Punic War, he designed war machines for the city's defenders. He died when the city was captured. In "Tuskulan Conversations," a Roman orator and statesman Mark Tullius Cicero (106–43 BC) wrote that in 75 BC, when he served as questor in Sicily, he was able to find in grave of a famous scientist in Syracuse. According to the will of Archimedes, a

Fig. 15 Archimedes. (Source: https://interesnyefakty.org/wp-content/uploads/arhimed-1.jpg)

ball inscribed in a cylinder was depicted on his grave. Later, the historian Plutarch (45–127) confirmed this.

Archipelago

Archipelago – is a complex of islands which are located close to each other. The geological structure of Archipelago is similar. According to their origin, there are the following types of Archipelago island: continental, volcanic, coral, alluvial. Small states may be situated upon the isles of Archipelago. They also may belong to a state closely located on the mainland. The islands are in contact with each other and have a geographical and political unity. Archipelagic lines are used for contact between the islands. All the water space inside these lines belongs to the archipelago-state. In the Mediterranean Sea there are 10 archipelagoes, among which the Ionian Islands Archipelago (q.v.).

"Archipelago"

"Archipelago" – is an Institute for the Marine Environment Protection, a nonprofit non-governmental organization, founded in 1998, Greece. "Archipelagos" conducts comprehensive research in the field of the protection of the marine environment in a wide region of the Greek seas and the northeastern Mediterranean. The activities of the institute integrate applied scientific research with environmental protection measures, which are held in close cooperation with local societies. This activity becomes a strategic framework that allows "Archipelagos" to act at the local, national, and European level to protect the rich biodiversity of the eastern Mediterranean region from many threats.

Arkoudi

Arkoudi – is an island located at 5.5 km south-east from the Syvota bay at the southern coast of Lefkas Island in the Ionian Sea, Greece. The area of the island is 4.2 km^2. The shape of the island is oval. Its length is 3 km, and width is 2 km. The banks are steep, rocky. The island's height is 143 m. The eastern part of the island is flat with cultivated fields and pastures for sheep and goat.

Association of Mediterranean Maritime Museums (*AMMM*)

Association of Mediterranean Maritime Museums (*AMMM*) – created as a nonprofit organization to maintain, enhance, and strengthen the activities of the maritime museums of the Mediterranean Basin. The association aims to guarantee the preservation of maritime history and culture in accordance with professional quality standards in relation to collections, heritage, ships, and projects; raising professional standards and supporting research related to maritime heritage and maritime history; ensuring and preserving maritime heritage; development of cooperation between members of the organization in the field of research publications, exhibitions, etc.; supporting the legislation favoring the realization of declared purposes. In addition, the association should work on the implementation of the aims declared by the International Congress of Maritime Museums (ICMM) and the International Council of Museums (ICOM), and act as an advisory committee in both organizations. The association is headed by a president and a vice president. Its work is carried out by the General Assembly and the Executive Committee in accordance with the existing Statutes. The headquarters of the association is located in the Maritime Museum of Barcelona, Spain.

Fig. 16 Atlantic bluefin tuna. (Source: https://krasivyblog.ru/wp-content/uploads/2021/03/tunets.jpg)

Astakos

Astakos – city and summer resort, which is located northeast of the Dragonera Islands (q.v.) above the bay of the same name in the area of Etoloakarnania, Greece. On the coast, a lot of small beaches with white sand: Aghios Yorgis, Asprogiali, Vela. Boats depart daily from Astakos to the islands of Ithaca and Kefalonia. Astakos was once a port, and an important one, which is evident from at least two parallel streets of the shopping district, which in the 1870s were built up with constructions in the classicism style.

Atlantic bluefin tuna (*Thunnus thynnus*)

Atlantic bluefin tuna (*Thunnus thynnus*) – a pelagic fish of the Scombridae family. It grows up to 4 m in length, reaching a weight of 1 t, swims at a speed of over 80 km/h. The most expensive fish in the world: In Tokyo, at the famous Tsukiji wholesale fish market, sushi lovers pay up to $280,000 per fish. Atlantic bluefin tuna meat is used to make popular sushi and sashimi. By the way, the muscle tissue of tuna fish contains a large amount of the amino acid histidine. In case of improper storage (violation of cooling or defrosting conditions), histidine under the impact of bacteria turns into histamine, which can cause poisoning. The problem is that histamine is quite resistant to high temperatures (Fig. 16).

The most acceptable conditions for tuna cooking are boiling, stewing, and hot smoking. It is better not to eat sushi with tuna. Atlantic bluefin tuna is common in the Mediterranean Sea and the Atlantic Ocean. Inhabits surface waters and depths up to 200 m. Maturation occurs at the age of 4–6 years. Spawning takes place in May–June in the western Mediterranean. Due to strong consumer demand, from 1997 to 2007, the abundance of redfin tuna in the East Atlantic and Mediterranean decreased by 60%. In subsequent years, it was already about 82%. Now there are just over 3 million of them. Moreover, 80% of the tuna mined is consumed in Japan. According to the results of numerous studies, scientists have made recommendations for a complete ban on fishing for this type of tuna in order to avoid its complete disappearance. In addition, it is recommended to expand aquaculture of this valuable fish through the establishment of tuna farms. The rescue plan for atlantic bluefin tuna is designed for 15 years and provides for the gradual reduction of quotas for its catch in the Atlantic, as well as the suspension of commercial fishing in the Mediterranean Sea. In September 2007, the European Union announced the closure of the tuna fishery, as all countries at that time had exhausted their quotas. However, the EU authorities reasonably suspected that their order was not being implemented, and the fishermen continued to fish and sell tuna to foreign buyers. Once again, the European Union has banned for an indefinite time the fishing of atlantic bluefin tuna in the Mediterranean Sea and the Atlantic since September 1, 2009. The ban on fishing was introduced at the request of the International Commission for the Conservation of Atlantic Tuna, since all quotas for commercial fishing introduced to restore the population were exhausted. In April 2009, the Principality of Monaco banned the sale and marketing of red tuna on its territory. This decision was facilitated by the Prince Albert Foundation and the World Wildlife Fund, in order to finally draw public attention to this problem. Monaco is the only country to have proposed in July 2009 to add atlantic bluefin tuna to the list of officially "endangered" species. Dozens of Japanese fishmongers have loudly voiced their opposition to plans to list atlantic bluefin tuna as threatened with extinction and therefore banned from trade. The European Union supported the idea of banning the international trade in Atlantic tuna at the CITES conference, which took place in March 2010 in Doha, Qatar. However, the proposal, under the aegis of the Principality of Monaco and with active US support, was widely discussed, but not accepted.

Atlantic bonito (*Sarda sarda*)

Atlantic bonito (*Sarda sarda*) – schooling predatory pelagic fish of the Scombridae family. The back is bluish-green, the sides and belly are silvery-white. On the upper half of the body there are usually at least 9 oblique longitudinal dark stripes. Length up to 90 cm (average 60–65 cm), weight up to 7 kg (usually 3–4 kg). Spawning is portioned. Fertility is 0.4–4 million eggs. Maturation is at 2 years. Growing fast. Juveniles feed on crustaceans, and adults on large fish (horse mackerel, mackerel, anchovies, sardines, etc.). It is distributed throughout the Mediterranean. Suitable for canning and smoking. The meat is characterized by a rich chemical composition. It contains chlorine and sodium, sulfur and zinc, manganese and fluorine, phosphorus, potassium, iodine, and other elements. In addition, fish contains a lot of vitamins (A, B, C, E, and PP). Regular consumption of this fish is believed to

Fig. 17 Atlantic bonito. (Source: http://fooduniversity.com/foodu/seafood_c/resources/ocean%20fin/Cold%20Water/Tuna/BonitoPacific.gif)

promote tissue regeneration and improve the overall health of the brain and spinal cord (Fig. 17).

Atlantic croaker (*Micropogonias undulatus, micropogon operculans Atlantic croaker, hard heads*)

Atlantic croaker (*Micropogonias undulatus, micropogon operculans Atlantic croaker, hard heads*) – fish of the Percidae family. The body is relatively high (in large individuals, humpbacked), laterally compressed. The head is large. The teeth are small, the jaw protrudes slightly forward. The dorsal fin is composed of two parts, separated by a notch. The second dorsal fin has a dark border at the point where the rays exit. The scales are relatively large and extend to the cheeks. Body coloration is silvery with dark stripes extending from the back to the sides. Typical sizes: length 25 cm, weight 500 g (maybe larger 1500 g). They dwell at a depth of 15–300 m, preferring silty, pebble and shell soils, hiding among reefs, and rocks. They reach sexual maturity in the second year of life. Females spawn 0.1–2 million eggs. Spawning lasts from August to October. The eggs are benthic, sinking to the soft muddy bottom, where the hatched larvae find enough benthos for feeding. The main diet of adult fish includes shrimp, crabs, various residues of plants and animals. The fish make croaking sounds that attract females by contracting and relaxing their swim bladder with special muscles. For that Atlantic croaker got its name – croaker (from the English "to croak"). It has an important commercial value. The fishing season is May–October. The meat is soft, with a delicate flavor (darker rather than white). (In Greek, it is called Pantelis, sykies.)

Atlantic Horse Mackerel, European Horse Mackerel (in the USA), Common Scad, Scad, and Saurel (Common Mackerel, Trachurus Trachurus)

Atlantic Horse Mackerel, European Horse Mackerel (in the USA), Common Scad, Scad, and Saurel (Common Mackerel, Trachurus Trachurus) – marine pelagic schooling fish of the Carangidae family. The back is dark, and the belly is silvery. There is a dark spot at the upper edge of the operculum. Body length is up to 40 cm, and weight is up to 1.8 kg. The fish of 15–25 cm long is usually caught. Keeps at has a bluish color with numerous transverse curved black stripes above the lateral line. The belly is bright silver. The length of the fish is up to 50 cm, and weight is up to 1.8 kg. It becomes sexually mature in the third year of life, performs long-distance migrations, and serves as food to predatory fish. The meat is delicious. In the eastern part of the Adriatic Sea catches reach 150 t (Fig. 18).

Atlantic Mackerel, Boston Mackerel, Norwegian Mackerel, Scottish Mackerel, or Mackerel (Scomber Scombrus)

Atlantic Mackerel, Boston Mackerel, Norwegian Mackerel, Scottish Mackerel, or Mackerel (Scomber Scombrus) – a pelagic schooling fish of Scomber genus, Scombridae family. The body is fusiform, slightly compressed from the sides, and entirely covered with fine scales.

Fig. 18 Atlantic horse mackerel. (Source: https://pbs.twimg.com/media/DlnaEIuU8AABHHq.png)

The head is pointed. The mouth is large, the eyes are large, and the teeth are small. Its back disguised in the ground. Its mouth and eyes are located on top, which creates the impression that it is constantly looking through the water at the sky (hence the name). It has small spikes on its broad head and operculum; the longest one, reverted, is located on the proscapula. The main color is graybrownish or gray-brown, dark on the back, and more lightful on the belly; on the sides there are dark longitudinal stripes, and the first dorsal fin is black. The length reaches 30 cm. It wages a sedentary lifestyle, and most of the time it lies buried in the sand, exposing only the eyes and half-open mouth, luring prey by a long, worm-like, mandibular respiratory membrane. Stargazer occurs in many areas of the Adriatic Sea, mainly at depths from 10 to 200 m. It breeds from May to August, fertility up to 125,000 eggs. Small fishes, such as bull-calves, goatfish, atherina, mollusks, and others, serve as food to it. Stargazer has no commercial value. It is also rare in amateur catches. To see it under water when it lies low is very difficult. When handling the stargazer, one must be careful since the punctures of its spines are very painful. Arts, the Technical University (Berlin, Germany), and the Worcester Polytechnic Institute (Boston, USA) (Fig. 19).

Fig. 19 Atlantic mackerel. (Source: https://i1.wp.com/simvol-shop.ru/wp-content/uploads/2021/03/skumbriya-scaled.jpg?fit=2560%2C1655&ssl=1)

Atokos

Atokos – an island located 9 km east of Ithaca island and separated from it by a deep-water strait of 8 km south-west of Kastos island, Greece. The westernmost of the Ionian Islands (Fig. 20).

The length of the island from north-north-east to south-south-west is 3.7 km, the greatest width in the southern part is 2.6 km. The area of the island is 4.4 km^2. The population is small in number. The height of the island is up to 30 m. The shores are high, mostly steep. Covered with bushes. In some places, there are small areas of cultivated land. Goats dwell here.

Augusta

Augusta – (1) city on Sicily island, Italy. Population 33,500 people; (2) the port, which is located in the bay of the same name, in the northern part of the vast bay of Ionian Sea, which forms an inlet into the eastern coast of Sicily Island between the capes of Sant'Elia and Santa Panagia. There is a naval base in the port. Oil is imported through the port of Augusta, paraffin, salt, oranges, and oil products are exported (Fig. 21).

In the autumn of 1908, a flotilla of Russian warships under the command of Admiral Litvinov — two battleships "Tsesarevich" and "Slava" and two cruisers "Admiral Makarov" and "Bogatyr," held on exercises in Augusta. On December 27, 1908, the ships shelled out joint gunnery drill, and at night the Messina earthquake occurred (q.v.). After this, the Russian flotilla set off from Augusta to Messina on a rescue mission; (3) a bay which consists of three parts: Xifonio bay, Megarese harbor, and Priolo bay. The bay of Xiphonio forms an inlet into the coast immediately northwest of Cape Izzo. There are quarries on its northeastern coast. Several piers protrude from this coast. The depth at the entrance to the bay is 10–20 m; toward the top, it gradually decreases. Megares harbor is located in the northwestern part of Augusta Bay. It is available for larger ships. It is protected by three breakwaters from the east and south. There are three forts in the harbor: Vittoria, Garcia, Avalos. The depths in the middle part of the harbor are 20–30 m. There are many piers and berths in it; (4) the island that separates the bay of Xiphonio from the harbor of Megares. It is located at 2 km west of Cape Izzo. The city of Augusta is located in its middle part.

Fig. 20 Atokos Island. (Source: https://upload.wikimedia.org/wikipedia/commons/thumb/3/38/Atokos.jpg/1600px-Atokos.jpg)

Fig. 21 Port of Augusta. (Source: https://sun9-57.userapi.com/impf/j1nc2jwSDdRDqZEKBbixTkvxvPFSSEgC7k-zqQ/_kG9HLvlSmQ.jpg?size=1280x960&quality=96&proxy=1&sign=0c8cfa39c3e7833c0039d8f53f8dcc15&c_uniq_tag=_CvshiN2KyD4YRX72lv98UMcpjfpOORAVm2wPDVFh38)

There is a fortress on the northern tip of the island. Across a strait of approximately 150 m, separating Augusta Island from the northern coast of Augusta Bay, two bridges are thrown. On the western coast of the island there is a large Torpediniere pier, 350 m long and 5.4–6.8 m deep, which belongs to the naval department.

Augustus (Gaius Octavius Until 44 BC, from 44 BC Gaius Julius Caesar Octavius, from 27 BC Gaius Julius Caesar Octavian Augustus) (63 BC–14 AD)

Augustus (Gaius Octavius Until 44 BC, from 44 BC Gaius Julius Caesar Octavius, from 27 BC Gaius Julius Caesar Octavian Augustus) (63 BC–14 AD) – Roman Emperor (from 27 BC), commander. After the victory over the combined Fleet of Mark Antony and the Egyptian queen Cleopatra (31 BC), he became an absolutist dictator. The Senate awarded him the honorary title of "Augustus," which means "Exalted by the gods." He founded a political regime, which was called the principate. During the Augustus reign, the conquest of Spain was accomplished. Then Egypt, Moesia, Pannonia, Dalmatia, and Germany were conquered. He continually improved the Army organization.

"Azov"

"Azov" – is a ship of the line, the best in the Russian Navy in the nineteenth century (Fig. 22).

It was built in 1826 in Arkhangelsk, Russia. The length of the ship was 52 m, and its width was 14 m, the depth indicator was 6 m. Its armament was: four 40-pound unicorn cannons; twenty-four 36-pound and thirty 24-pound cannons; twenty-two 24-pound carronades. In October 1827 it took part in the Navarino naval battle (q.v.) under the command of M.P. Lazarev, captain I rank, as part of the United Allied Squadron (Russia, France, England). He destroyed five Turkish corvettes with cannon fire, including the frigate of the Turkish-Egyptian fleet commander, Ibrahim Pasha. The ship itself got 153 hits, seven of them

Fig. 22 Azov. (Source: https://a.d-cd.net/M4AAAgHhsOA-960.jpg)

below the waterline. The ship was completely repaired and restored by March 1828. The highest award was given to Azov: for the first time in the history of the Russian fleet she was granted by the "Admiral St. George ensign and pennant in honor of the commendable acts of commanders, courage and fearlessness of officers and men." After the "Azov" was excluded from the lists of the Russian fleet in 1831 its Order Flag passed to the ship "Memory of Azov," named after the heroic predecessor, and when in 1848 this ship was also excluded from the lists of the fleet, the flag was raised on a new ship with the same name. When the second ship "Memory of Azov" was decommissioned, the flag was handed over to the museum. The Order Flag of Azov is kept in the Hermitage. The area of this flag is more than 120 m^2 (14.5 × 8.5 m). M.P. Lazarev for his heroic deeds was granted by the title of Rear Admiral. Young future Russian naval commanders Lieutenant P.S. Nakhimov, midshipman V.A. Kornilov, and garde-marine V.I. Istomin were obtained the higher ranks and bemedalled.

B

Baillie Henry (About 1763–1826)

Baillie Henry (About 1763–1826) – Rear Admiral. In 1783, he was taken from the English Fleet to serve in the Russian Fleet as a Midshipman and sent to the Don Flotilla. In 1784–1787 he commanded a small ship in the Sea of Azov as a Lieutenant. In 1788, as a commander of the schooner "Pobedoslava Dunaysky," cruised with the Fleet and participated in the battle of Fidonisi. In 1789 he was promoted to Captain-Lieutenant and conducted a campaign on the Black Sea. In 1790 he was in a cruise and participated in the battles of the Kerch Strait and Hajibey, and in 1791 at the Kaliakra Cape. In 1792–1798 he commanded the frigate on the Black Sea. In 1798, Baillie went with the archipelago squadron of F.F. Ushakov on the Mediterranean Sea, and participated in the liberation of the islands of Cerigo, Zante, in the blockade of Corfu. In 1799, Baillie was promoted to Captain 2 rank, commander of the frigate "Schastliviy." He commanded the landing operations at the city of Manfredonia, Italy, from where he began a victorious march to Naples, captured the city of Foggia, and occupied Naples. When Emperor Paul I learned about this, he declared: "Baillie surprised me, and I will surprise him too!" He awarded the Lieutenant-commander with the General's Order of St. Anna of I class. In 1800–1801 he was at the shores of Naples. In 1802 he returned to Sevastopol. In 1803 he was promoted to Captain of the first rank. In 1804, he commanded the ship "Azov" in the squadron of Admiral D.N. Senyavin, moved from Sevastopol to Corfu, and the next year he cruised in the Mediterranean. In 1806, with a detachment of three ships, he arrived from Corfu in the Adriatic Sea and captured the Port of Cattaro (Kotor) in Montenegro, which became the base for the Russian Mediterranean squadron, as well as the Curzalo fortress and Lissa Island. In 1807, he sailed in the Adriatic Sea, participated in the blockade of Patras Island, and was inducted into the rank of Captain Commander. In 1808–1812 (as an Englishman, since there was a war with England) by the highest order he was sent to Saratov. In 1812 he returned and was appointed to command the ship on the Black Sea, and in 1813 at the Port of Sevastopol. In the years 1814–1816 he commanded the 59th Naval crew in Sevastopol. In 1816 he was promoted to Rear Admiral. In 1817–1826 commanded the 3rd Naval brigade in Sevastopol (Fig. 1).

Balkan peninsula

Balkan Peninsula – the peninsula in southern Europe. The area is about 505,00 km^2. The greatest distance from west to east is approximately 1260 km, from north to south – 950 km. It is washed from the west by the Adriatic Sea and the Ionian Sea, from the east the Black, Marmara, and Aegean seas. Conventional continental

I. S. Zonn, *The Ionian Sea Encyclopedia*, Encyclopedia of Seas,
https://doi.org/10.1007/978-3-031-08206-1

Fig. 1 Baillie Henry (*Belly Henry*). (Source: http://www.tri-veka.ru/upload/iblock/a18/488.jpg)

borders of the Balkan Peninsula are the Danube River to the mouth of the Sava River, and then the line to the top of the Gulf of Trieste of the Adriatic Sea. Bulgaria, the most of the territory that was previously part of the state of Yugoslavia, Albania, Greece, the southern regions of Romania, the western regions of Turkey, and a small northeastern region of Italy are situated on the Balkan Peninsula. The shores are heavily dissected. They are mostly high, steep, with small bays of a Dalmatian type in the west and lobed type in the south. The coastline of the Adriatic Sea is extremely embayed. There are also a large number of islands, especially in the Aegean Sea. The total area of islands adjacent to the Balkan Peninsula is about 21,500 km^2. The topography of the Balkan Peninsula is predominantly mountainous. It is characterized by the extreme roughness of the surface, and the presence of numerous intermontane depressions. In the northeast of the peninsula there are the Pirin Mountains, Rila (Musala, 2925 m, is the highest mountain in the Balkan Peninsula), and the Rhodope Mountains. To the north of the Rhodope Mountains stretch the Stara Planina Mountains, separated from them by longitudinal hollows. In the north they gradually fall to the hilly Danube Plain. These mountains are characterized by the predominance of relatively gentle slopes and peaks of soft outlines. In Rila and Pirin occur alpine type the topography. Along the entire western margin of the Balkan Peninsula stretches the Dinaric Plateau to the south in the mountains of Pind and the mountains of the Peloponnese Peninsula. These mountains are characterized by steep slopes and an abundance of deep gorges; they are characterized by plateau-like summits. Karst topography forms are widely developed, especially in the west and north-west of the Dinaric Highland (Karst Plateau or Kras). The plains are mainly related to marginal areas, and partly to intramountain depressions. The largest of them are the Thracian, Thessaloniki, Thessalian, Albanian Lowlands, the southern parts of the Lower Danube, and Middle Danube Plains (Fig. 2). The climate in the central and northern regions of the Balkan Peninsula is moderately continental, and in the rest part of the peninsula it is Mediterranean. In addition to these basic types, there are various transitional climates. In the mountains, there is a high-altitude climate zone. The average July temperature in the north is 22–23 °C, and in the south 25–27 °C; the average January temperature, respectively, is from −1 °C to 2 °C (in the mountains it decreases below −5 °C) to 8–11 °C. In the mountainous west, the annual precipitation is 1000–1500 mm; in the eastern and southern part of the Balkan Peninsula (except for high mountains) the precipitation is less than 1000 mm; and in some places it is less than 500 mm. The longest rivers are the Maritsa, Morava, Iskar, and Struma and the border Danube and Sava Rivers. Most of the rivers are mountainous, cataracted, tumultuous (especially along the upper reaches); they are often alternate with parts of the mountain and plain flows. The rivers are deep in winter and spring. Low water occurs in summer. Small rivers in the southern and eastern parts of the peninsula dry up in summer. Rivers are characterized by energy value, they are used for irrigation, and the largest ones are suitable for shipping. The large lakes are Skadar, Ohrid, and Prespa which are located in tectonic depressions. There are many karst lakes. Lakes in the Rila Mountains are of glacial origin. There are many crescentic lakes in the valley of the Danube

Fig. 2 Balkan Peninsula. (Source: https://i.pinimg.com/originals/ac/aa/e1/acaae11060870362e6385de45d906ddf.jpg)

River. Subtropical brown and mountain brown typical and calcareous soils predominate in the southern part of the peninsula; on the Adriatic coast, the red terrarosa soils are widespread. In the mountainous and foothill regions of the northern part of the Balkan Peninsula, there are mountain-forest brown and mountain-forest humus-calcareous, and brown forest soils are developed. On the Danube Plain chernozems and transitional types from brown forest soils to brown soils are widely spread. The black compacted Smolnitsa soils occur on the Maritsa River Lowland. There are more than 6500 species of plants on the Balkan Peninsula. There is a large percentage of endemic species (about 27%). The slopes of the mountains are often rocky and bare. On the south in the lower mountainous belt zone there is Mediterranean vegetation with evergreen oak and pine forests and tangles of shrubs (maquis). Above 400–500 m, as well as in the areas located further to the north, there is a transitional belt with forests of evergreen and deciduous broadleaf trees; tangles of xerophytic deciduous shrubs (shibljak) are specific for this area. In the north and in the mountains of the inner part of the Balkan Peninsula, the forests formed of Central European species prevail: in the lower zone, there is mainly deciduous oak, and in the higher areas, and there are beech and coniferous (fir and pine) species. The upper boundary of the forest is located at an altitude of 1800–2000 m. On the plains of the extreme northeast, steppe vegetation occurs. Steppe and forest-steppe plains, the northern and eastern parts are plowed up; corn, wheat, tobacco, and grapes are cultivated; there are also plum and apple orchards. In addition, in the south olives, citrus fruits, and pomegranates are cultivated. On the most part of the Balkan Peninsula farming requires artificial irrigation. Among mammals there are fox, otter, wildcat, wild boar, red deer, roe deer, chamois, bear, jackal, Iberian lynx, mole rat, and bats. The fauna of birds, reptiles (lizards, snakes, Greek tortoises), amphibians (tritons, frogs, toads, in the western part of the caves – olms), and mollusks are rich and diverse. Albania, Bosnia and Herzegovina, Bulgaria, Croatia, Greece, Kosovo, Macedonia, Montenegro, Romania, Serbia, Slovenia, and Turkey are situated on the Balkan Peninsula. Greece, Romania, Bulgaria, Slovenia, and Croatia are members of the European Union. Albania, Bulgaria, Croatia, Greece, Montenegro, Romania, Slovenia, and Turkey are members of NATO. On the territory of the Balkan Peninsula, there are important transport routes connecting Western Europe with South-West Asia. The Balkan Peninsula is home to many ancient cultures and civilizations.

Balkan Union

Balkan Union – military-political union of Bulgaria, Serbia, Greece, and Montenegro. The union was created in 1912 under the auspices of the Russian Empire against Austria-Hungary. The Balkan Union began a feud with the Ottoman Empire, since a large number of Bulgarians, Greeks, and Serbs lived in the weakening empire. In addition, the Bulgarian government wanted to extend the borders of Bulgaria, creating a Greater Bulgaria – an empire that was supposed to cover the entire eastern part of the Balkans. The Serbs desired access to the Adriatic Sea, annexing Western Macedonia and Albania to their country. Montenegrins sought to occupy the large Turkish ports on the Adriatic and Novi Pazar sandjak. The Greeks, as well as the Bulgarians, needed to extend the boundaries of their country. There were, however, contradictions in the union. Thus, Greece, Bulgaria, and Serbia argued about the identity of Macedonia, and Greece and Bulgaria – about the identity of Thrace.

Basento

Basento – river in the south of Italy. It flows through the territory of the Basilicata region (q.v.) to the east, and flows into the Gulf of Taranto in the Ionian Sea. Near the city of Metaponto. The length is 149 km, and the basin area is 1537 km^2. Average consumption is 12.2 m^3/s. It originates in the Lucan Apennines, Mount Arizo, southwest of Potenza. In the valley of Basento are the cities of Potenza, Tricarico, Ferrandina, are Metaponto.

Basilicata

Basilicata – an administrative region of southern Italy, located at the Gulf of Taranto in the Ionian Sea. Its area is 9995 km^2. The population is 547,579 inhabitants (2020). It borders with Apulia in the north, with Apulia and Ionian Sea in the east, with Calabria in the south, and the Tyrrhenian Sea and Campania in the west. The capital is Potenza, the highest regional capital that has survived many earthquakes and natural disasters. The region is divided into two provinces – Potenza and Matera. Forty-seven percent of the territory of Basilicata is covered with mountains. Partially, the Pollino massif enters here, the highest point of which is Dolcedorme (2267 m). The Sirino massif with the heights of Monte Papa (2004 m), Monte Raparo (1764 m), and the Maddalena mountain system with the Monte Volturino peak (1836 m) also pass here. The Spento volcano is located in the north of the province. 8% of the territory is covered with plains (Fig. 3).

Most of them are confined to the sea coast. Of the rivers, the most significant are Bradano, Basento, Agri, Sinni, and Cavone. The climate is subtropical Mediterranean. On the coast, the climate is typically marine; the further inland, the more continental it becomes. Basilicata is one of the Italian regions rich in fauna and flora, and there are many different natural parks. Basilicata has been inhabited by people since ancient times, as evidenced by archaeological data. This is the only region of Italy that has two names – Basilicata and Lucania. There are different assumptions about the origin of both names. One of them is from the Greek word "basileos," that is, the title of the Byzantine emperor. The name appeared during the period when Italy was under the rule of

Fig. 3 Administrative region of Basilicata. (Source: https://www.italynews.ru/files/10_Basilicata_Mappa.jpg)

Byzantium. This name was regularized and is now official. However, there were two periods in history when the region returned to the previous, more ancient name – Lucania. This was from 1799 to 1820 and from 1932 to 1947. This name comes from the Lucan people, whose name is explained by the Latin "Lucus" (translated as forest) or the Greek "Lykos" (wolf). In the eighth century BC Greek colonists landed on the coast of the Ionian Sea. From this time, a new historical period begins for the south of Italy, which was noted in the annals as Great Greece. The Lucans, along with the Samnites and Bruttians, as the inhabitants of Calabria were called in those times, resisted both the Greek colonialists and the Roman conquerors. Ultimately, a stronger people – the Latins – conquered all of Italy. After the establishment of Roman rule in the southern regions of Italy, a period of desolation and poverty began. In the Middle Ages, Basilicata was subjected to numerous raids by Germanic tribes, Goths (Visigoths and Ostrogoths), and Lombards, after which it was under the rule of Byzantium. In 847 Basilicata became part of the autonomous principality of Salerno. Since the eleventh century Basilicata became part of the Kingdom of the Two Sicilies, and then the Kingdom of Naples (one of the two parts of the previous one). In these kingdoms, two dynasties fought for power – Anjou and Aragonese. Throughout its history, Basilicata remained the poorest region of Italy, which gave rise to constant peasant uprisings, revolts, robbery ("brigantaggio"), and mass migration. The latter is relevant to this day. This is the poorest region of Italy. Cereals are presented as the main ones in the agriculture – wheat, corn, barley, and oats. Basilicata is the country's leader in the production of these crops. On the hills there are vineyards, and fruit plantations, where plums, peaches, pears, kiwi, and citrus fruits are grown.

Sheep, pigs, and goats are mainly presented in cattle breeding. Fishing is underdeveloped (only on the coast of the Ionian Sea). Basilicata's basic product is olive oil. Cheese is made from sheep and goat milk. Red peppers and the famous Lucan sausage (lukanika) from pork are also traditional here. The popular wine Aglianico del Vulture is also produced in the region. Tourism is poorly developed; however, thanks to this, clean and comfortable beaches stretch along the coast.

Battle of Lepanto, 1571

Battle of Lepanto, 1571 – naval battle in the Gulf of Patras at Cape Scrofa between the fleets of the Holy League and the Ottoman Empire (Fig. 4).

The forces of the united Holy League represented the strongest and most numerous European fleet. It included about 300 different vessels, of which 108 Venetian galleys, 81 Spanish galleys, 32 papal galleys, and other Italian states, as well as 6 huge Venetian galleases (q.v.). The number of ship crews was about 84 thsd. People, of which about 20 thsd. Were soldiers from boarding teams. Although sometimes other data are given. The Turkish fleet consisted of about 210 galleys and 66 galiots. The number of teams and boarding squads could reach 88 thsd. People. (of which about 16 thsd. in boarding teams). Ali Pasha-Müezzinzade commanded the Turkish fleet. The allied fleet blockaded Turkish ships in Gulf of Patras. On the morning of October 7, 1571, both fleets met at the entrance to the bay 60 km from the city of Lepanto (now Nafpaktos). The battle formation of the Turkish fleet consisted of the center, two wings, and a small reserve located behind the center (5 galleys, 25 galiots). The right wing of the Turks (53 galleys, 3 galiots), headed by Mahomet Sirocco, was the weakest. The center division (91 galleys, 5 galiots) was commanded by Ali Pasha himself. The left wing (61 galleys, 32 galiots) consisted mainly of the ships of the Algerian pirates, headed by Uluç Ali. The forces of the Turkish fleet stretched

Fig. 4 Battle of Lepanto. (Source: http://www.vestnik-evropy.ru/issues/stseny-s-mesta-kazni-ili-kaznit-nelzya-pomilovat-pesa-perevod-s-angliyskogo-aleksandra-sergievskogo.html)

for 8–10 km. The Allied fleet formed in the same order of battle. The center division (62 galleys) was headed by don Juan de Austria. The right wing (58 galleys) was commanded by Giovanni Andrea Doria (great-nephew of the famous Admiral A. Doria). The left wing of the Allies (53 galleys) was led into battle by the Venetian Barbarigo.

Thirty galleys under the command of the Marquess of Santa Cruz were assigned to the reserve. Don Juan gave orders to unleash the Christian rowers and arm them. Both fleets moved forward. After the contact between the fleets, three skirmishes broke out simultaneously. The left wing of the Allies, due to lack of knowledge of the terrain and fear of taking the ground, kept away from the coast. The Turks took advantage of this.

Galleys from the first wing outflanked the allies along the coast and attacked from the rear. Part of the Turkish galleys wedged between the center of the enemy and his left wing. As a result, the entire left wing of the Christians was surrounded. Barbarigo was forced to take a boarding battle surrounded, and here the advantage of the allies in armament and the number of boarding teams affected. In the afternoon, the Turks who surrounded Barbarigo were defeated. The encirclement of the enemy ships did not give the Turks any benefit, since in close combat the enemy turned out to be much stronger. In the center, where the main forces of the opponents clashed, the battle was stubborn. The main targets of the attack were the flagship galleys of don Juan de Austria ("Real") and Ali Pasha's ("Sultana"). Ultimately Ali Pasha was killed in a shootout. The center of the Turks began to give their ground and retreat. The commander of the left wing of the Turkish fleet, Uluç Ali, with most of his wing turned to the center and struck from the side at the forces of Juan de Austria. The flagship galley of Ali Pasha was destroyed, and Juan, breaking the general order, began to turn toward the ships of Uluç Ali. At the same time, the Allied reserve under the command of the Marquis of Santa Cruz entered the battle. The commander of the right flank of the allies, Doria, also turned and began to approach the center of the allied battle formation, directly on Uluç Ali. Uluç Ali's ships could be surrounded, thus, he began to withdraw from the battle. The defeat of the Turkish fleet was complete, and historians differ only in the assessment of losses.

The following numbers are most often called: the Turks lost 224 ships, including 117 were captured by the allies. Twelve thousand slaves were captured and freed on Turkish ships. At least 10 thousand slave rowers (Christians, since Muslims could not be kept in slavery) died along with the sunken ships. Up to 15 thousand Turkish soldiers and sailors were slained. According to various estimates, from 300 to 5000 Turks were captured. Thirty Turkish guns were also captured. The losses of the allies were much lower. It is believed that the Allies lost 13 galleys, 7500 men slained and 8000 wounded. It should be noted that the future famous Spanish writer Miguel Cervantes (q.v.) (the author of "Don Quixote of La Mancha") participated in the Battle of Lepanto and got three gunshot wounds, after which his hand was inactive for all his life. In "Don Quixote" he wrote about Leopanto that this is "the greatest of events that have ever happened in the past and in the present, and are unlikely to occur in the coming century." The battle was the largest naval battle of the sixteenth century, showing that the Turks, invincible until that time, could be defeated. The battle is reflected in European culture. It was embedded in the paintings of F. Bertelli (1572), and H. Luna (1887), P. Veronese (1572). Despite a brilliant tactical victory for the Holy League, the battle had little impact on the overall course of the war. Taking advantage of the lack of unity among the allies, Turkey quickly created a new fleet and successfully ended the war.

Under the peace treaty of 1573, Venice ceded to the Ottoman Empire Cyprus Island and pledged to pay a substantial contribution.

Battle of Modon, the Second Battle of Lepanto (*Modon Battle*)

Battle of Modon, the Second Battle of Lepanto (*Modon Battle*) – the naval battle that took place in

August 1500 during the Venetian-Ottoman War of 1499–1503. In December 1499, the Venetians attacked Lepanto, hoping to regain the territories lost after the defeat at the Battle of Zonchio (q.v.). Kemal Reis sailed from Cefalonia and recaptured Lepanto from the Venetians. He stayed in Lepanto (q.v.) between April and May 1500, where his ships were repaired by an army of 15,000 craftsmen who were brought from the area. Then he sailed from there and fired at the Venetian port on the island of Corfu, and in August 1500 again defeated the Venetian fleet, after which he captured the Modon fortress by attack from the sea. He then intercepted the Venetian fleet near Coron and took the city along with the Venetian brigantine in the harbor. From there he sailed towards the Island of Sapienza (now – Sapientza), where he sank the Venetian galley "Lezza." In September 1500 Kemal Reis assaulted Voiussa, and in October 1500 appeared at Cape Santa Maria on the island of Lefkada. In November, he returned to Istanbul (Constantinople), ending the campaign.

Battle of Navarino, 1827

Battle of Navarino, 1827 – a major historic naval battle between the united squadron of Russia, England, and France, on the one hand, and the Turkish-Egyptian fleet, on the other. It happened on October 8 (20), 1827, in the Navarino Bay of the Ionian Sea on the southwestern coast of the Peloponnese Peninsula (Fig. 5).

The battle is one of the episodes of the Greek National Liberation Revolution of 1821–1829. In 1827, the London Convention was signed between England, France and Russia, according to which Greece was granted full autonomy. The Ottoman Empire refused to recognize the convention. Precisely in 1827, the combined squadron of Russia, Great Britain, and France (1276 guns in total) under the command of a senior in the rank of British Vice-Admiral Edward Codrington (q.v.) approached Navarino Bay, where the Turkish-Egyptian fleet was located (up to 2200 guns in total) under the command of Moharram Bey. Ibrahim Pasha was the commander-in-chief of the Turkish-Egyptian troops and fleet. Ibrahim Pasha's fleet consisted of 66 battle units, including three ships of the line, 23 frigates, 40 corvettes and brigs. In addition, the Turkish-Egyptian fleet supported 165 coastal guns and six fire ships. The allied squadron consisted of 26 ships, including 10 battleships, 10 frigates, and six corvettes and brigs (1250 guns). The allies were inferior in artillery, but superior in combat training of personnel. Codrington hoped by a demonstration of

Fig. 5 Battle of Navarino. (Source: https://ae01.alicdn.com/kf/HTB13FZzKpXXXXaDXXXXq6xXFXXXD/Navarino.jpg_q50.jpg)

force (without the use of arms) to force the enemy to accept the demands of the Allies. Thus, for this purpose, he sent a squadron to the Navarino Bay. The united fleet approached the entrance of Navarino harbor in two columns. One of them consisted of English and French ships, and the other of a Russian squadron. After the Allied convoy passed the fortress batteries and anchored, the Russian convoy with the common anchor ship "Azov" in front (q.v.) approached the harbor entrance. According to the disposition announced by Codrington, the ships of the united squadron were located in a semicircle: English and French on the flanks, Russian – in the center. The core of the Russian detachment was the four best battleships of the Russian fleet: "Azov," "Gangut," "Iezekiil,"and "Aleksandr Nevskii." Each of them had to fight a heavy battle with several enemy ships simultaneously, and they all fought heroically. At this time on one of the Turkish fireships there was a strong gunfire, in which the English lieutenant FitzRoy, who was sent as envoy, was slain. His task was to force the fireship commander to move further from the allied ships. After a while, the first shot in the direction of the French frigate was heard from one of the Egyptian corvettes. The battle began after the Turks killed the second envoy, who was sent to the flagship of Moharram Bey. It lasted about 4 h and ended with the destruction of the Turkish-Egyptian fleet. Russian squadron under the command of Rear Admiral L.P. Geyden was the most decisive and skillful. The squadron was defeated the entire center and right flank of the enemy fleet. It took the major blow of the enemy and destroyed most of his ships. In the battle, the flagship "Azov" distinguished herself (it got 153 hits, of which seven were below the waterline), for the first time in the Russian fleet "Azov" was awarded a special Flag of St. George. Under the command of M.P. Lazarev, the future Russian naval commanders P.S. Nakhimov, V.A. Kornilov, and V.I. Istomin approved themselves. The casualties of the Turkish-Egyptian fleet amounted to more than 60 ships and more than 4000 people killed and wounded. The allies did not lose a single ship. Their losses in killed and wounded were 600–800 people. After the battle, the allied fleet remained in Navarino Bay till October 26. The defeat of the Turkish fleet in the Battle of Navarino significantly weakened the naval forces of Turkey, which served as a significant contribution to the victory of Russia in the further Russo-Turkish war of 1828–1829. The Battle of Navarino provided support for the Greek national liberation movement, which, according to the Treaty of Adrianople of 1829, resulted in the autonomy of Greece. In Greece, in the town of Pylos, a monument to three naval commanders-admirals of the Navarino battle was erected. By the 100th anniversary of the battle, stamps were issued with the image of admirals, and a monument to the slain sailors of Russia, France, and England was erected. In Oranienbaum (now – Lomonosov, St. Petersburg) in the former home of Admiral Anjou, Navarin Day is celebrated annually on October 8.

Battle of Patras (1772) (*Patraikos Battle 1772*)

Battle of Patras (1772) (*Patraikos Battle 1772*) – naval battle that took place on October 26–29 (November 6–9, new style) in 1772 during the Russo-Turkish war of 1768–1774 in the Patras (Patraikos) Gulf off the coast of Greece. After the defeat of the main forces of the Turkish fleet in the Battle of Chesma and active operations of the Russian fleet in the Archipelago during 1770–1771 remained the most significant "Dulcinite" squadron, named after its base – the city of Dulcigno (modern Ulcinj in Montenegro), consisting of 47 frigates and xebecs with artillery – 16–30 guns, with transports, which were up to 8 thsd. Men. In October 1772, the "Dulcinite" squadron left Dulcigno and headed for Navarino, where it hoped to take on ships from the coastal fortresses of Modon, Koroni, and Navarino up to 4 thsd. Men. In order to bolster the fleet operating in the archipelago, on May 1772 the Russian government sent a fourth squadron under the command of Captain first Rank M.T. Konyaev. Captain Konyaev with his squadron cruised off the coast of the island of Kythira, where on October

Fig. 6 Battle of Patras. (Source: https://upload.wikimedia.org/wikipedia/commons/3/30/Jacob_Philipp_Hackert_Patras.jpg)

16, 1772, he met with Major Voinovich's squadron, who conveyed to Konyaev the order to advance against the "Dulcinite" squadron of the Turks. When he learned that Kapudan Pasha (Admiral) Mustafa Pasha with his fleet is in the Gulf of Patras and awaiting 12 more ships with a landing from Corfu, Konyaev decided to attack the Turks immediately. At the time of approaching Patras, Konyaev had at his disposal two battleships ("Graf Orlov" – 64 guns, "Tshesma" – 74 guns), 2 frigates ("Sv. Nikolai" – 26 guns, "Slava" – 16 guns), 2 "polaccas" ("Modon" and "Ausa" – 12 guns each), and one xebec ("Zabiaka" – 18 guns). In total, Russian ships had only 224 guns. The Turks had 9 frigates (30 guns each) and 16 xebecs (some had 30, others had 20 guns). In total, the Turkish squadron had up to 630 guns. Surprise and swiftness, command skill, and outstanding bravery ensured the victory of the Russian fleet in this battle. At the same time, the Russians had almost no casualties at all. The Russian squadron burned seven frigates and eight xebecs. One frigate managed to be pulled into the Gulf of Lepanto (Gulf of Corinth), but was already so damaged that it sank the next day. Six xebecs managed to flee. All Russian ships survived. In the Russian crews, the losses were absolutely negligible: one killed and seven wounded. The Turks had losses of 200 people. The main result of the battle was a very significant weakening of the "Dulcinite" squadron. After this defeat, the Turks never again bothered the Russian fleet based in the Archipelago until the very end of the war (Fig. 6).

Battle of Preveza

Battle of Preveza – naval battle between the forces of the Ottoman Empire navy and the coalition of the Holy League fleets (Spain, Venice, Genoa, Papal States), which took place on September 28, 1538, at the Ambracian Gulf (Gulf of Arta or the Gulf of Actium) near Preveza in the Ionian Sea in northwestern Greece. In 1537, the Ottoman corsair Hayreddin Barbarossa captured and plundered many islands in the Aegean Sea belonging to Venice, and annexed the Duchy of Naxos to the Ottoman Empire. These actions forced Pope Paul III to unite several Christian states in the Holy League to fight the Turkish fleet. The allied fleet gathered near the island of Kerkyra, and the Genoese Admiral Andrea Doria became its commander-in-chief. The coalition had a numerical superiority over the Turkish fleet: their forces consisted of 157 galleys, while the Turks had only 122. At dawn on September 28, the Turkish fleet left Preveza and headed south, where the coalition fleet was anchored at the island of Sessola, near the western coast of island of Lefkada. The battle ended in a crushing defeat for the European fleet: the Turks sank 10, burned 3 and captured 36 ships and 3000 prisoners.

Battle of Zonchio, Battle of Sapienza or the First Battle of Lepanto (*Zonkio Battle*)

Battle of Zonchio, Battle of Sapienza or the First Battle of Lepanto (*Zonkio Battle*) – naval battle during the Venetian-Ottoman war of 1499–1503, which took place on four separate days: 12, 20, 22, and 25 August 1499. The first naval battle in history using cannons mounted on ships. In January 1499, the Turkish admiral Kemal-Reis set sail from Istanbul (Constantinople) with 10 galleys and four ships of other types, and in July met the huge Ottoman fleet sent to him by Davud Pasha in order to wage a large-scale war against the Venetian Republic. He commanded 67 galleys, 20 galiots, and about 200 smaller vessels. Having reached Cape Zonchio in the Ionian Sea in August 1499, between the Greek cities of Pylos and Methoni, the fleet under the command of Kemal Reis defeated the Venetian fleet of 47 galleys, 17 Galiots, and about 100 smaller vessels commanded by Antonio Grimani. At the most critical moment of the battle, two Venetian carracks captained by Andrea Loredan and Alban d'Armer boarded one of the flagships of the Turkish fleet. The commander of the Turkish battleship Burak Reis was unable to disentangle his ship from the boarders and chose to set her aflame. The sight of the three great ships burning together dealt a severe blow to the Venetian morale.

Black seabream (*Spondyliosoma cantharus*)

Black seabream (*Spondyliosoma cantharus*) – rather deep bodied fish of the Sparidae family with a relatively small head. The jaws reach the anterior edge of the eye from behind. The teeth are sharp, thinning, slightly crooked, and not very large. The size of the teeth and their shape in the front of the jaw and on its sides are the same, only the front teeth are slightly larger. The top is grayish, and the sides are silverish-gray with six or seven vertical dark stripes. The fins are also grayish, and the rim of the caudal fin is dark. As an exception, it reaches 51 cm in length, 3 kg in weight, in most cases about 35 cm, and in catches it is about 0.5 kg. Black Seabream is a fairly common fish in the seas of Northern Europe, but appears in the waters north of the English Channel only in summer. A typical inhabitant of rocky areas, although in the Mediterranean Sea it is usually found on a sandy or seaweed bottom. It is one of the few Sparidae fish that build nests. The male digs holes on the sandy bottom, in which the female lays eggs. The male protects the eggs until hatching; juveniles stay in the nest area for several weeks after hatching. In some areas, amateur fishermen catch it and it is even an object of fishing; however, its number is not large enough to serve as the basis for intensive fishing. About 460 tons of this fish are caught in the Mediterranean Sea.

Blue flag

"Blue Flag" – an international symbol award. Since 1987 it has been annually awarded to beaches and marinas that meet high environmental standards and have high-quality safety and service facilities. The "Blue Flag" is a free certificate of beaches quality, a symbol of high standards of service quality maintained by them for a long time, which is issued only for 1 year. Prize candidates usually have to be approved at the national level, and then the final decision is made by the European jury. France is considered the birthplace of the "Blue Flag," where the first award ceremony took place in 1985. The "Blue Flag" nomination is managed by the Foundation for Environmental Education (FEE), which is a nonprofit nongovernmental organization. The fund consists of 65 organizations located in 60 countries of the world and representing Europe, Africa, Oceania, Asia, North and South America.

Fig. 7 Beach, awarded by "Blue Flag". (Source: https://www.topcrop.ru/wp-content/uploads/2016/07/flag.jpg)

Eligibility criteria for the "Blue Flag" require observing strict water quality, safety, and environmental education standards. For places located in the EU countries, full eligibility for the adopted directives is also necessary. The annual award ceremony takes place on June 5 (for countries of Europe, Canada, Morocco, Tunisia, and similar geographical location) and on November 1 (for countries of the Caribbean, New Zealand, South Africa, and the others like them, located in the southern hemisphere). All countries located on the coast of the Adriatic Sea, except Albania and Bosnia-Herzegovina, participate in the program "Blue Flag." At the beginning of 2020 there are more than 100 Adriatic blue flag beaches in Italy, 9 in Slovenia, more than 100 in Croatia, and 31 beaches in Montenegro. Currently, 47 countries of the world participate in the "Blue Flag" program (Fig. 7).

Blue shark (*Prionace glauca*)

Blue shark (*Prionace glauca*) – a species of cartilaginous fish from the Carcharhinidae family. It is one of the most common sharks on Earth. This species is specific to subtropical and moderately warm waters. It lives at depths from 0 to 350 m. It has an elongated slim body with long pectoral fins. Body color on the top is blue, on the sides it becomes brighter, and the belly is white. The shark reaches 3.8 m in length and weighs 200 kg. The maximum recorded weight was 391 kg. The first dorsal fin begins at the caudal margin of the pectoral fins. The ridge between dorsal fins is absent. The small growths occur at the tailset. The teeth are triangular, beveled, with jags, but without lateral teeth. The lower ones are significantly different from the upper ones: they are less massive and not always jagged. Usually Blue sharks hunt on bony fish, mollusks such as squids, cuttlefish, and octopuses, and crustaceans. Also, its diet includes small sharks, mammals' corpses, and sometimes sea birds. The remains of whales and common porpoises were found in the stomachs of the caught Blue sharks. It is known that Blue sharks eat cod caught in trawl nets. They rarely hunt tunas. Adolescent Blue sharks do not have natural enemies in nature, except humans. Juveniles may fall prey to large sharks. This is a viviparous species of sharks. A placental junction is formed from the yolk sac between the mother and the embryo. The number of baby sharks in litter varies from 4 to 135 with a size of about 40 cm. Pregnancy lasts for 9–12 months. Females reach sexual maturity at the age of 5–6 years, and males 4–5. During the prelude to mating, the male

Fig. 8 Blue shark. (Source: https://tourjournal.ru/wp-content/uploads/Golubaya-sinyaya-akula.jpg)

bites the female, so the sex of sharks can be easily identified by the presence or absence of scars on the back. The skin on the back of females is three times thicker than that of males. Blue sharks' meat is used in food; however, it is not very popular. On the world market, the meat of this shark sometimes comes under the following names, for example, "gray fish," "stone salmon," and "sea eel." A high content of heavy metals (mercury, lead) has been reported in the meat of Blue sharks. In addition, a fish meal is made of the Blue shark, fins are used for soup, and vitamins are made of liver fat (Fig. 8).

Bluefish (*Pomatomus saltatrix*)

Bluefish (*Pomatomus saltatrix*) – schooling pelagic fish of the Pomatomidae family. The head is oval, the mouth is large, and the teeth are large and sharp. The back is greenish or greenish-blue, and the belly is silvery. It can reach a length of up to 1 m and a weight of 10 kg, usually 20–50 cm. It occurs throughout the Mediterranean Sea. Performs summer migrations. Its roe is pelagic. Fertility is 100 thousand to one million eggs. The incubation period is 2 days. A gluttonous predator that consumes large quantities of sardine, mackerel, and anchovy. According to FAO statistics, annual catches in the Mediterranean reach approximately 15,000 (Fig. 9).

Bora, bura

Bora, bura – cold gusty wind from the north and north-east. Observed in the area of the southeastern coast of Italy. Sometimes its speed reaches 20 m/s or more. If Bora lasts 2–3 days, then there also may be destruction, and communication with the shore becomes difficult. Quite often Bora's duration reaches 3–7 days.

Borino

Borino – summer winds from the northeast. Usually they are weaker than bora winds; however, sometimes they achieve great strength. Borino, as a rule, does not last longer than 3 days and blows under a clear sky.

Bridge Lefkada (Lefkas Bridge)

Bridge Lefkada (Lefkas Bridge) – a freely pontoon floating road bridge through which Lefkada

Fig. 9 Bluefish. (Source: https://adriaticnature.com/wp-content/uploads/2016/02/Pomatomus-saltatrix.jpeg)

connects to mainland Greece. It was built in the twentieth century. Its width is 12 m, length is about 30 m. Every hour it is raised in order to allow ships to pass along the strait. The crossing capacity is quite high – up to 600 cars per hour. In addition, there are two walking paths. In 1864, Lefkada, along with all the Ionian Islands, became part of Greece. The fortification is located opposite the bridge.

Butrint (*Butrinti*)

Butrint (*Butrinti*) – 1) a national park with an area of 29 km^2, an archaeological museum-reserve in the south of Albania near the city of Sarandë, near the border with Greece. Located on the shore of the lake of the same name, 2 km from the Ionian Sea, opposite Kerkyra Island. A modern Albanian village called Butrint is located nearby. Butrint is the former capital of Epirus, a small kingdom located in the southern part of the Balkan Peninsula. The kingdom became famous in the whole world because of one ruler – Pyrrhus, who, fighting against the Romans, won a victory, which cost him a huge price and the loss of many fellow tribesmen. The bleeding army was disgracefully defeated in the next battle against the Romans. Hence the expression "Pyrrhic victory" came from. The ancient city of Butrint was founded by Greek colonists from Corfu Island, directly opposite the entrance to the sea bay of the same name, in the sixth century BC local residents traditionally were engaged in fish farming in the freshwater gulf, traded food products with residents of the opposite island Kerkyra (Corfu Island). The city grew rapidly and became rich, however, the Roman Empire fell and was destroyed in 476 by vandals. Its western part fell under the rule of the Ostrogoths, whose king – Theodoric – did everything in his power to preserve Roman culture. However, long wars destroyed the roads – the engine of progress and trade in the Roman Empire, and the city of Bitrunti began to decline. The settlement managed to be under the rule of the Bulgarian Kingdom, and from the IX century became part of the Byzantine Empire. In the fourteenth century Ottoman invaders came here. They tried to spread Islam on the territory of the Balkan Peninsula and next to it in Greece, Bulgaria, and Serbia. Already in the fifteenth century Bitrunti came under the control of the Venetian Republic. A fortress at the top of a high hill has been preserved from the time of its reign (Fig. 10).

In its basement there is a museum, where there are interesting archaeological finds made during the excavations of Bitrunti. This museum is considered the most valuable museum in Albania, which houses many unique examples of the country's history. Bitrunti was partially excavated in 1928–1935 by the Italian scientist L. Ugolini. The massive walls with gates (including "Lion's Gate with a relief image of a killing bull") of the fifth to sixth centuries BC that surrounded the acropolis were revealed. In the lower city the sanctuary of Asclepius with a statue of the god, theater of the third century BC with 19 rows of benches and marble statues, the remains of

Fig. 10 Bitrunti National Park. (Source: https://mybalkans.ru/wp-content/uploads/2017/09/theatre-ruins-unesco-world-heritage-site-butrint-albania.jpg)

residential and public buildings were discovered. After World War II, excavations continued. Currently, the Butrint Foundation, founded by the British Lord Rothschild, supports this unique territory in every possible way. In 1992, the excavations of Greek and Roman cities were included in the UNESCO World Heritage List. In 1999 the site was expanded to include a medieval Venetian fortress; 2) lake, tectonic lagoon on the coast of the Ionian Sea in the south of Albania. It is separated from the Ionian Sea by the Ksamil Peninsula. It is connected with the Corfu Strait of the Ionian Sea by the Vivari channel 3600 m long, 60–100 m wide, and 5–6 m deep. The surface area of the lake is 16 km^2, the drainage area is 128 km^2, and salinity is 20–35%. The exchange of sea waters occurs every 6 h with a slight (15–20 cm) tide. A small amount of fresh water comes to the northern part of the lake from the Bistritsa and Kalos rivers. The southern part of the lake is part of the Butrint National Park.

Byron George Gordon (*George Gordon Byron*) (*1788–1824*)

Byron George Gordon (*George Gordon Byron*) (*1788–1824*) – one of the greatest English romantic poets of "broken heart poetry," the national hero of Greece. Born in London. At the age of ten, he inherited the title of baron from his great-uncle and the family estate of Newsted Abbey near Nottingham, where he moved with his mother. Studied with a home teacher, then he was sent to a private school in Dulwich, and in 1801 – in Harrow. In the fall of 1805, Byron entered Trinity College, Cambridge University. In 1806, Byron published exclusive book "Fugitive Pieces," and a year later - "Hours of Idleness." Fleeing from creditors, and probably also in search of new impressions, Byron in 1809 went on a long journey: Lisbon, (Portugal) Spain, Albania, and Athens (Greece). There he spent the winter in the house of a widow, whose daughter, Teresa Makri, sang in the image of the Maid of Athens. In the spring of 1809, on his way to Constantinople, Byron crossed the Dardanelles, which he was later proud of. In 1811, Byron returned to England, bringing with him a manuscript of an autobiographical poem, telling about the grieved vagabond. "Child Harold's Pilgrimage," published in 1812, glorified the name of Byron. In 1812, Byron spoke in the House of Lords with his first speech – against the Tory bill on the death penalty for weavers, deliberately breaking recently invented knitting machines (Fig. 11).

The success of "Childe Harold" provided a warm welcome in society for Byron. In the

Fig. 11 G.G. Byron. (Source: https://rgnp.ru/wp-content/uploads/5/0/2/502f830c674f0c8454a9cb7f2e1f2d83.jpeg)

footsteps of "Childe Harold" Byron created a cycle of "Oriental poems": "The Giaour" and "The Bride of Abydos" – in 1813, "The Corsair" and "Lara" – in 1814. At the beginning of 1815, Byron married A. Milbenk, niece of Lady Melbourne. However, the marriage was unsuccessful. Suspicions of incest and homosexual relationships and constant brief love affairs – debauchery before marriage – were confirmed. In 1816 he sailed to Europe, where he settled in Geneva. Here Byron completed the third song "Childe Harold," wrote "The Prisoner of Chillon" and began "Manfred." In September 1816, Byron went to Italy. In Venice, Byron studied the Armenian language, attended the theater of the Countess Albrizzi and her salon, and finished "Manfred," a drama in verse on a Faustian theme. Here he writes the fourth canto of "Childe Harold." Byron returned to Venice in November, having already written "Beppo" – a brilliant, ironic satire on Venetian customs. In the middle of 1817 he began a new satire called "Don Juan." He sold his family estate Newsted in the fall of 1818 and got rid of debts. In June 1819, Byron went to Ravenna, and then to Venice, and again to Ravenna in January 1820. His residency in Ravenna was extremely fruitful for Byron. He wrote new cantos for "Don Juan," "The Prophecy of Dante," a historical drama in verse "Marino Faliero," and translated the poem "Big Morgante" by L. Pulci. During the autumn and winter of 1820, he actively participated in the conspiracy of the Carbonari, members of a secret political movement against Austrian tyranny. In the midst of the conspiracy, Byron created a drama in verse "Sardanapalus" – about an idle voluptuary, who is prompted by circumstances to a noble deed. Before leaving Ravenna (in October) Byron wrote his most evil and unusual satire "The Vision of Judgment," and also finished the drama in verse "Cain," which embodied his skeptical interpretation of Biblical plots. In September 1822, Byron moved to Genoa, where he returned to work on "Don Juan," and by May 1823 he completed the 16th canto. Byron consistently leads Juan through a series of adventures, sometimes funny, then touching, – from the "platonic" seduction of a hero in Spain to idyllic love on a Greek island, from a slave state in a harem to the position of the favorite of Catherine the Great, and leaves him entangled in the networks of love intrigue in an English country mansion. Byron cherished an ambitious plan to bring his roguish novel in verse to 50, if not more, cantos, however, managed to finish only 16 and 14 stanzas of song 17. In 1823 the London Greek Committee offered Byron to help Greece in the war of independence. He agrees and lives for about 4 months on the island of Kefalonia (q.v.), awaiting instructions from the Committee. Byron liked to sit on the hill, which is located between the capital of Kefalonia Island – Argostoli – and the fortress of St. George near the village of Lakithra. Here he wrote: "If I am a poet, then I owe this to the air of Greece." Byron gave money to equip the Greek fleet and at the beginning of 1824 he joined Prince Mavrokordatos in Missolonghi. He took under his command a detachment of Suliotes (Greco-Albanians), to whom he paid a monetary allowance. Here, having fallen ill with a fever, he died. There is a monument dedicated to him in the park of heroes. In addition, a monument to Byron is installed in Athens and on Corfu Island.

C

Cachalot (*Physeter macrocephalus*)

Cachalot (Sperm Whale) (*Physeter macrocephalus*) – marine mammal, the only modern member of the sperm whale family (Physeteridae). The largest of the toothed whales. Adult males reach 20 m in length and weigh 50 tons, and females 15 m in length and weigh 20 tons. This is one of the few cetaceans in which females are much smaller than males, reaching a length of only 15 m, and also differ in constitution, number of teeth, head size, shape, etc. (Fig. 1).

The whale has a huge rectangular head, which holds the so-called spermaceti organ. The sperm whale is a gregarious animal that lives in large groups, sometimes reaching hundreds and even thousands of heads. Distributed throughout the World Ocean, except the polar regions. It feeds mainly on cephalopods, including giant squid, reaching a length of 14–18 m. Fish usually contains no more than 5% of its diet. In search of prey, the whale makes the deepest dives among all marine mammals – to a depth of over 2 km, remaining under water for up to one and a half hours. The sperm whale searches for prey using ultrasonic echolocation, in which a spermacetic organ, used as an acoustic lens, plays an important role. The spermacetic organ also helps provide the whale with the required buoyancy during diving. In nature, the sperm whale has practically no enemies. However, man has been hunting the sperm whale for a long time – in the past, this whale was the most important object of whaling. Due to over-hunting, which ceased only by the 1980s, the number of sperm whales has greatly decreased. At the present, it is slowly regenerating, although this is somewhat hindered by anthropogenic factors (pollution of the seas, intensive fishing, etc.). It is hard to estimate the number of sperm whales in the world; the most likely number is 300–400 thousand heads. The sperm whale is considered a very dangerous animal. When wounded, it is very aggressive, and therefore hunting it was associated with a serious risk. The enraged sperm whales killed many whaling sailors and even sunk several whaling ships. The sperm whale is considered the "pearl" of the Ionian and Aegean Seas. It is also found in the south of the Tyrrhenian Sea and is very rare in the Adriatic Sea.

Caïque

Caïque (*Kaik, Kayik*) – (1) a small fishing vessel common in the Middle East, Ionian and Mediterranean seas. It has an elongated forward end, round bilge, and a high stern. Sailing rig - sprint rig, rectangular gall topsail, forestaysail, and one or more jibs.

(2) A small light rowing (occasionally sailing) vessel. Used mainly in the calm waters of the Bosphorus, Turkey. One of the luxurious caïques was presented to the Russian Emperor Nicholas

I. S. Zonn, *The Ionian Sea Encyclopedia*, Encyclopedia of Seas,
https://doi.org/10.1007/978-3-031-08206-1

Fig. 1 Cachalot (Sperm Whale). (Source: https://web-zoopark.ru/wp-content/uploads/2018/06/7-159.jpg)

I by the Turkish Sultan. Several of these Sultan's caïques are preserved in the Istanbul Maritime Museum.

(3) A small boat used in Greece for island-to-island travel or sailing by sea. Frequently chartered by travel agents.

Calabria

Calabria – a peninsula in southern Italy, the southern part of the Apennine Peninsula between the Ionian Sea, the Strait of Messina, and the Tyrrhenian Sea. Length about 250 km, width 40–100 km. Most of the surface is occupied by the Calabrian Apennines, composed in the east mainly by limestones and flysh, in the west by granites and gneisses. Height 1950 m (in the Aspromonte massif). On the slopes of the mountains, maquis, in some places beech and pine forests, and along the coast narrow, highly hilly, cultivated plains. The territory of Calabria is prone to earthquakes. Major cities are Reggio di Calabria, Catanzaro, and Cosenza (Fig. 2).

Fig. 2 Calabria. (Source https://mirsg.ru/shared/upload/IMAGES/Directions/Yacht/Sicilia/catania.jpg)

Calypso deep

Calypso deep – located in the Ionian Sea southwest of the island of Pylos, Greece (36°34′N, 21°8′E). The deepest part of the Mediterranean Sea with a minimum depth of 5267 m. In the deep, the African Plate slides beneath the Aegean Sea Plate, creating the Hellenic Trench.

Cape Pelorus

Cape Pelorus (Faro Point (Italian Punta del Faro)) (*Peloro*) – a promontory located at the northeastern tip of the island of Sicily, Italy. Low and sandy. The shores of the cape are eroded, as a result of which its configuration varies. To the southwest of the cape is the Torre Faro village, where the pilot station is situated. The marine pilots of this station have the right of piloting ships throughout the Strait of Messina. Two lakes, Pantano Piccolo and Pantano Grande, are located 1 km west of the promontory, connected by a canal. The Peloro lighthouse is installed on the cape. Forte Spuria is located on a wooded hill with a height of 100 m, 2.7 km west of the cape, which houses a storm signal station.

Capo d'otranto lighthouse

Capo d'otranto lighthouse – a lighthouse located in a place in eastern Italy, where the waters of the Ionian and Adriatic seas meet. Built in 1867. It stopped operating in the 1970s. It was subsequently restored and opened to tourists since 2008 as a museum object. It is a tall stone structure 32 m high. The lighthouse tower emerges from the caretaker's two-story house. The entire structure is dazzling white.

Capo Passero

Capo Passero – the extreme southeastern island of the Ionian Sea, 21 m high, Italy. The extreme south of the Sicily island is separated from it by a shallow strait of 550 m wide. It is devoid of vegetation. The south-eastern coast is steep and rocky with grottoes and cliffs, and the northwestern coast is gentle, and there is a beautiful beach of fine sand here. At the northeastern end there is an ancient castle, the Castello Fortezza of Emperor Charles V. There is also the Passero lighthouse. The depths in this part of the sea are 16–18 m. To the west of the island, at an altitude of 25 m, lies the village of Capo Passero.

Caretta caretta sea turtle

Caretta caretta sea turtle – q.v. loggerhead sea turtle.

Cassibile

Cassibile (*Cassibeli*) – river in the southeast of Sicily Island, Italy. Length 30 km. The source is in the Hyblaean Mountains, not far from Palazzolo Acreide. It flows to the southeast and flows into the Ionian Sea, 23 km south of Syracuse. There are several waterfalls and canyons on the river. The canyon area is a special area: Riserva naturale orientata Cavagrande del Cassibile.

Catania

Catania – (1) one of the three territorial units of the same name of Sicily, Italy. It is located on the east coast of the island of Sicily, including the Ionian coast, the plains, and Mount Etna. It borders in the south with the provinces of Syracuse, Ragusa, in the north with Messina, in the west Enna, Caltanissetta. The area is 3553 km^2. The population is 1,084,000 people (2008). A significant part of the province's territory is occupied by the fertile Plain of Catania – the largest in Sicily. The northwest of the province is occupied by mountain ranges, the largest of which is Mount Etna. The largest river basin of Sicily is located on the territory of the province: Simeto river with tributaries –

Gornalunga and Dittaino rivers. The total area of this basin is over 4300 km^2. It goes beyond the province. The province is washed by the Ionian Sea for 65 km. All types of beaches (sandy, pebble and mixed) are found here. Some parts of the coast are occupied by outcrops of basalt rocks. Agriculture is reduced mainly to the cultivation of citrus crops and viticulture, with a significant proportion of the grapes being table varieties. In terms of industrial development, the province ranks first in Sicily. The food industry, winemaking, alcoholic beverage production, chemical, pharmaceutical, instrument-making, and electronic industries, as well as the production of building materials, fishing, agriculture, information technology, and tourism prevail. The capital of the province is the city of the same name. Catania is linked to other Sicilian provinces by bus, rail, sea, and air connections. Around Mount Etna there is the local "Circumetnea" railway.

(2) A bay on the east coast of the Sicily Island on the Ionian Sea. It forms an inlet into the coast between Cape Molini and Cape Campolato, 31 km south of it. The western coast of the bay is flat, low, bordered by a sandy beach. The Simeto River flows from this bank into the bay. To the north of its mouth lies a fertile plain, and to the south is hilly and in some places boggy area. The southern coast is steep, bordered by rocks. A railway runs along it. The gulf forms an inlet into the coast for 9 km, the length of its coastline is about 39 km. In the north, its coast is rocky and impregnable. It consists of rocks, mainly basalt, which, as a result of repeated eruptions of Mount Etna, descended to the sea. Some of the rocks, protruding from the water, form a small archipelago of islets near Sicily. On the largest of these islands is one of the main attractions of Catania – Aci Castello (Castello Normano). In the south of the bay (behind the breakwater and the harbor) there is a beautiful sandy beach. There are several caves in the bay that are inaccessible for intromission from land, as well as a large port and several fishing villages.

(3) The city of Sicily, the second largest on the coast of the Ionian Sea on the southern side of the active volcano Etna. The population is 311,500 people (2019). The administrative center of the province bears the same name. Founded in 729 BC by Greek colonists who arrived here from the Euboean city of Chalkis. A native of the city was its semi-legendary legislator Charondas, whose laws were of extreme severity, however, and were adopted in other Chalcidian colonies in Italy and Sicily under the name "Chalcis regulations." In 476 BC the city was conquered by Hieron I – a cruel Syracuse tyrant who expelled the inhabitants and renamed the city with the name Etna. After the death of the tyrant, people moved back and returned Catania its original name. However, the period of independence was short: in 403 BC another ruler of Syracuse subjugated the city and sold its inhabitants into slavery. The archaeological monument of this era is the statue of Cora (Persephone) with a torch. For almost 150 years, Catania was dependent on Syracuse. In 263 BC during the first Punic War, the city was captured by the Romans and up to the fifth century was part of the Roman Empire. These several centuries were a period of flourishing economy and culture. A forum was built here (the ruins of which have been preserved near the courtyard of San Pantaleone), a circus for chariot competitions, a theater and an odeon, and an amphitheater in the northern part. The creation of a water supply system is dated by the same period, which indicates a high standard of living of the townspeople. After the fall of Rome, the city was ruled by the Byzantine Empire (Fig. 3).

In 827 the Arabs landed in Sicily, and later, in 1071, they were driven out by the Normans led by Roger I. In 1169 Catania was destroyed by a strong earthquake. Thirty years later, the city authorities supported Hautevilles in the fight against Henry VI – the son of Frederick Barbarossa. In response, the emperor's troops captured and burned part of the city, including the cathedral. During the reign of Frederick II, Catania finally freed itself from feudal dependence. In 1347, as a result of the pestilence epidemy, the city was practically depopulated. The foundation of the first university here in Sicily (1434) can be considered a confirmation of the cultural status of Catania. Second half of the seventeenth century was a tragic period in the history of the city: in 1669 there was a strong eruption of Mount Etna, most of the houses were buried under a layer of lava and ash. In 1693, while Catania had

Fig. 3 City of Catania. (Source: https://maps.spravka-region.ru/img/1/6338977428005057.jpg)

not yet recovered from the first blow, an earthquake destroyed everything that was preserved or was built. However, the inhabitants did not leave the city and rebuilt it again. Black lava and light limestone were used as the main materials. In the nineteenth century. Catania rebelled against the Bourbons in 1837 and 1848, and later became part of the United Italy. From this period the city has been developing rapidly. Only during World War II its development slightly decelerated. Over the past few decades, high-tech manufacturing has been created in Catania, which is why it is sometimes called the European Silicon Valley. Catania is an important transport hub. There are ship repair yards, a motive-power depot, agricultural engineering enterprises, electronic, oil refining, and chemical industries. There is a gas field near the city. There is a subway (1 line, stations are mainly of ground type). Catania's "Fontanarossa" Airport is located 5 km from the city and bears the name of the famous Catania native Vincenzo Bellini. The seaport of Catania is situated at 2 km from the city. In addition, not far from Catania, there is the port of Etna, famous throughout Sicily. There is also a developed Marina di Riposto, which can accommodate 360 yachts with a side length of up to 40 m. Symbols of the city are u Liotru, or the Fontana dell'Elefante, Odeon of the III century, the Roman Forum, the Castello Ursino of the thirteenth century, the Botanical Garden.

(4) A port located in an artificial harbor created du to the construction of a breakwater almost 2 km long. The resulting water space between the coastline and the concrete breakwater has become a reliable harbor for merchant, passenger, and leisure boats. The first port was built on the initiative of King Alphonse of Aragon in the twelfth century. However, severe storms several times tore down the piers of the moorings and threw moored ships ashore. The most dramatic event took place in 1601, when a severe storm turned all the coastal structures into a heap of stones. Only at the beginning of the eighteenth century, life in the port began to revive, and by the second half of the nineteenth century, it turned into a busy transport hub, where the transshipment of goods onto the

railroad became possible. Today, in addition to merchant and cargo ships, passenger ferries are moored in the port, on which you can get to Genoa, Civitavecchia, Naples, and Malta.

Catanzaro

Catanzaro – a city located on a wide rocky promontory jutting into the northern part of the Gulf of Squillace of the Ionian Sea, Italy. The population is 91,000 people. Founded by the Byzantines in the tenth century in the ancient heart of Calabria – the city of Sila (Fig. 4).

An active agricultural and commercial center with developed industrial and craft activities. Traces of the past are hidden in its artistic treasures. Attractions include: Villa Trieste (also known as Villa Margherita), a city garden with a beautiful view of the Ionian Sea and the Lesser Sila mountains, the Church of St. Dominic (between the XV and XVI centuries), the Chiesa dell'Osservanza (XV century), a provincial museum displaying prehistoric materials of various origins, as well as archaeological exhibits from different eras.

Catanzaro di marina

Catanzaro di marina – a resort town which is situated in the Gulf of Squillace of the Ionian Sea south of Catanzaro, Calabria province, Italy. An Italian resort famous for its simplicity. The "face" of the city is its broad sandy beaches, which are equipped with everything necessary for a good rest. The ruins of an ancient Roman polis, ancient signal towers, the ruins of the San Rabano Abbey, the Solidarity center, the Maremma Natural Park are interesting.

Celes

Celes – Greek pirate ship used in the Mediterranean in the fourth to first centuries BC. It was a small, fast galley with a straight bow, one row of rowing benches and one or more sails. Despite the

Fig. 4 Catanzaro. (Source: https://img-fotki.yandex.ru/get/484029/98350100.5c/0_c5f3d_417d581f_XXL.jpg)

small hold, the Greek pirates used such craft, preferring them due to their speed.

Cephalonia or Kefalonia

Cephalonia or Kefalonia (*Kefallinia or Kephallenia*) – an island which is situated at the mouth of Gulf of Patras, between the islands of Zakynthos and Lefkada, the farthest from the mainland. The capital of the island is Argostoli. Its area is 781 km^2, the length of the coastline is 254 km, and the population is 35,800 people. (2011). The island is a satellite of the Ithaca island. There is a narrow strait 3.5 km wide between them. The largest and most mountainous island in the Ionian archipelago (Fig. 5).

The Ainos mountain range reaches an altitude of 1628 m above sea level. Since 1962 it has been a national reserve. Its slopes are covered with snow in winter. Unique Kefalonian black spruce, from which Odysseus built ships, which have not been preserved anywhere else and are included in the Red Book. Small valleys are located in the areas of Omala, Sami, Pilaros, Livadi, and Paliki. Drinking water supplies are concentrated in the Omala and Pyrgi regions, where the large Avithos lake was formed. The coastline of the island is heavily indented. The largest bays are Sami, Myrtos, Argostoli, Lourdas, and Livadi. The topography of the coastline in the east is different than in the west. From the side of the mainland, the mountains slope down gently to the valleys, smoothly merging with the edge of the sea, while the eastern side is severely bristling with reefs. Natural caves Melissani, Aggalaki, St. Theodore, Zervati, Drogaratis, and others are of particular interest. According to archeological data, Kefalonia was inhabited during the Paleolithic era. The ancient Greek tribe of Leleges inhabited it from the fifteenth century BC. Homer mentions Kefalonia under the name Sami. Certainly, one of the most beautiful is Myrtos Beach with its emerald blue sea, white sand, and small pebbles. It is also worth stopping by the red sand beach in Xi Bay, next to which is another miracle of Kefalonia – the rhythmically swaying rock of Kounopetra. Guidebooks call it heaven on earth. In 2006, "Forbes" magazine singled out Myrtos among European beaches as one of the cleanest and most beautiful in Europe. The photo of this beach has become the hallmark of the island. Today's Kefalonia, except for Fiskardo, has been rebuilt according to very strict rules – all buildings, especially schools, hospitals, and hotels, are earthquake resistant. The new buildings retain the style of old Kefalonia. The climate of Kefalonia is typical of the Mediterranean climate zone: hot and dry, sunny summers, mild and rainy winters. Among the attractions – Katavotres. Nearby there is a memorial dedicated to the Italian soldiers who were executed by the German fascists during World War II. In the south of the island, during the excavations, the ruins of a Roman villa were discovered, among which were found well-preserved mosaics

Fig. 5 Kefalonia Island. (Source: https://www.marinareservation.com/articles/wp-content/uploads/2018/12/kefalonia_marinas.jpg)

and inscriptions of the third century AD. The Monastery of Agios Andreas has been preserved since the time of Byzantium; its church contains magnificent frescoes and a church museum where the right foot of the Holy Apostle Andrew is kept. The Castle of Saint George is a Venetian fortress, the first capital of the island from 1500 to 1757. It was built as a polyhedron with three bastions. There is the Monastery of Agios (Saint) Gerasimos, where the Venerable protector of the island lived as a anachorite. His incorruptible reliquiae are preserved, to which believers come to worship and pray. Drogaratis is a cave with rare stalactites and stalagmites.

Concerts are held in a huge underground hall with unique acoustics. Melissani is an underground cave-lake with magnificent stalactites, in ancient times – the sanctuary of the god Pan. Part of the cave ceiling fell into the lake and formed a magical island. At the hour when the sun's rays penetrate through the ceiling hole into the cave, the water sparkles in all shades of blue and dark blue. The British submarine "Perseus," which was blown up by a mine in 1941 rests near the coast of Kefalonia.

Cer

Cer – the strait separating the island of Kerkyra from the western coast of the mainland – the Balkan Peninsula. The shores of the strait, except its middle part, are mountainous. The middle part of the eastern bank is a vast plain along which the Thyamis river flows. The shores are overgrown with vast vegetation. Shrubs and trees cover the high cliffs of the mountain ranges. In the valleys and on the slopes of the mountains – cultivated fields and plantations of olive trees. The eastern and western shores of the strait are heavily indented. The depth of the strait is 55–70 m. As you approach the shores, the depths decrease slightly. In winter, south-east and south-west winds dominate in the strait, accompanied by low cloud cover. Strong northern and northwestern winds sometimes blow, however, they are short-termed. In summer there are light breezes. The current in the strait, as a rule, is directed to the north and reaches high speed in its narrowest part. Surface currents reach a speed of 1.5–2 knots.

Cerigo

Cerigo – the old name of Kythira island.

Cervantes Miguel de Saavedra (1547–1616)

Cervantes Miguel de Saavedra (1547–1616) – world famous Spanish writer, author of one of the greatest works of world literature – the novel "The Ingenious Gentleman Don Quixote of La Mancha." The date of his birth is September 29, 1547 (the day of Saint Michael the Archangel). It was discovered approximately on the basis of the records of the church book and also on the tradition of giving a child a name in honor of a saint whose holiday fell on a child's birthday. It is certainly known that Cervantes was baptized on October 9, 1547, in the church of Alcalá de Henares. Some biographers claim that Cervantes studied at the University of Salamanca, but there is no evidence for this version. There is also a version that he studied with the Jesuits in Cordoba or Seville. Then he left Castile and went to Italy, where he got acquainted with the art of the Renaissance. By 1570, Cervantes was enlisted as a soldier in the Marine Corps regiment of Spain located in Naples (Fig. 6).

He stayed there for about a year before starting active service. In September 1571, Cervantes sailed aboard the "Marquesa," which was part of the Holy League galley fleet, which on October 7 defeated the Ottoman flotilla at the Battle of Lepanto in the Gulf of Patras in the Ionian Sea. Cervantes fought bravely aboard the ship and received three gunshot wounds – two in the chest and one in the forearm, which rendered his left arm mobile. In his poem "Journey to Parnassus" he wrote: "lost the movement of the left hand

Fig. 6 Cervantes Miguel de Saavedra. (Source: https://interesnyefakty.org/wp-content/uploads/Migel-Servantes-3.jpg)

for the glory of the right" (he thought about the success of the first part of Don Quixote). Cervantes proudly recalled his participation in this battle: he believed that he had taken part in an event that would determine the course of European history. After the Battle of Lepanto, from 1572 to 1575 Cervantes continued his service, being mainly in Naples. In addition, he participated in expeditions to Corfu and Navarino, and witnessed the capture of Tunis and La Goulette by the Turks in 1574. Cervantes was in Portugal, also made trips on service to Oran (1580s), and served in Seville. In September 1575, when returning from Naples to Barcelona aboard the "Sol" galley, the galley was attacked by Algerian corsairs on its way to the Catalan coast. Cervantes was imprisoned and taken to Algeria. In Algerian prison, Cervantes spent 5 years (1575–1580), tried to escape four times and only miraculously was not executed. At the end of 1580 he was redeemed from captivity for 2000 ducats. Cervantes's literary career began when he was 38 years old. The first novel, "La Galatea" (1585), was followed by a large number of dramatic plays with little success. In search of earnings, he enters the quartermaster service. He is assigned to purchase provisions for the "Invincible Armada," then appointed as a collector of arrears. During the performance of these duties, Cervantes goes to prison on charges of embezzlement. His life in those years was a whole chain of severe asperities, hardships, and calamities. From 1598 to 1603 he appears in Valladolid, where he is engaged in small private affairs. In 1604, the first part of the novel "The Ingenious Gentleman Don Quixote of La Mancha" was published, which had tremendous success in Spain and abroad (translated into many languages). From that time until his death, Cervantes's literary activity did not cease: in the interval between 1604 and 1616, the second part of "Don Quixote" was published, all the short stories, many dramatic works ("The Jealous Extremaduran," "Theatre of Miracles," "The Labyrinth of Love" and others), the poem "Journey to Parnassus" was written. The novel "Persiles and Sigismunda" was published after the author's death. The memory of Cervantes is immortalized in monuments erected in Madrid, Toledo, Seville, Moscow, and in the Greek city of Nafpaktos (formerly Lepanto). He entered the top five immortal Spaniards – Cervantes, El Cid, El Greco, Velazquez, and Don Quixote.

Charadros

Charadros – a river in the northern part of Achaea, Greece. The length is about 25 km. It originates in the mountains of Panachaiko. It flows in a northwesterly direction through a narrow forested valley. In its lower course it passes through the villages of Skioessa and Kato Kastritsi. The University of Patras is situated on its north bank. It flows into the Gulf of Patras of the Ionian Sea.

Charybdis

Charybdis – (1) a mythical sea monster that lived opposite Scylla in the Strait of Messina between Italy and Sicily. Three times a day, it absorbed the water and exhaled it back with a terrifying roar. Odysseus managed to pass Charybdis during his journey.

(2) The maelstrom formed by the Montante current occurs near the Faro village. Sometimes there are tide races that are dangerous for small vessels.

Chelura

Chelura – the marine wood-boring crustacean, which is larger than Limnotoria. Usually it settles next to it and drills the wood similarly to it, but its passages are deeper, although it does not create "cauldrons." The diameter of the moves reaches 2.5 mm.

Chrisanthi

Chrisanthi (*Hrisanti*) – a channel 2400 m long and 42.7 m wide, dug in the shallow southern part of the Preveza Strait in the north-east-south-west direction. The depth of the channel is 7.3 m (1951).

Chub Mackerel (Pacific ChubMackerel, *Scomber japonicus*)

Chub Mackerel (Pacific Chub Mackerel, *Scomber japonicus*) – pelagic fish of Scombridae family. It refers to fast-growing species of fish. By the end of the first year of life, it reaches a length of 18–20 cm. In commercial catches, it is found to be from 16 to 50 cm long. Body weight varies from 60 to 1200 g. It reaches sexual maturity with a length of 27–30 cm in the third year, and in the mass in the fourth year of life. The optimum water temperature for spawning is 14–21 °C, and depth is up to 250 m on spawning grounds. Japanese mackerel is characterized by extended migrations with a speed of 100–200 to 350–450 miles per month with a total length of 500–1500 miles. The ability to overcome significant distances is provided by the ability of mackerel to use the movement of water masses for traveling. In all cases, the vertical distribution is limited by the temperature jump layer and varies from 20–50 m in the summer to 200–250 m in the winter. Due to the absence of a swimming bladder in mackerel, the fish can quickly and painlessly make vertical movements in the water column. It is excellent fish for table use. The meat is juicy and fatty.

Ciane

Ciane – river in Italy. It is situated in the south of the island of Sicily. The length of the river is 37 km. The source is in the mountains of Hybalean Mountains. It flows into the Ionian Sea near Syracuse. The power of the river flow varies greatly depending on the season. In summer, the river can practically dry up, however, in winter, during rains, it can overflow.

City Aquarium

City Aquarium is situated in the city of Messina, Italy, near the coast of the Ionian Sea. Here, in eight large aquariums, the flora and fauna of several specific areas of the Mediterranean are presented.

Codrington Edward (1770–1851)

Codrington Edward (1770–1851) – sir, British admiral. In 1783, 13-year-old Cordington was enlisted in the Royal Navy and joined the yacht "Augusta," then he served on the 44-gun frigate "Assistant," 50-gun ship "Leander," 32-gun frigate "Ambusquede" and ships "Formidable" and "Queen Charlotte." In 1793 Cordington was promoted to the rank of lieutenant and assigned to the ship "Santa Margarita," then transferred to the 28-gun corvette "Pegasus." In 1794 he served on the 100-gun ship "Queen Charlotte." Cordington distinguished himself in the battle of the British squadron at Ouessan. He was promoted to the rank of commander and appointed commander of the "Kometa" fireboat. A year later he was promoted to the rank of captain and appointed commander of the sloop "Babet." Participated in the Battle of Groix Island. In 1796 he cruised off the coast of Portugal. From 1797 to 1803 he was on the coast. In 1803, after the end of the Treaty of Amiens and the resumption of hostilities, Cordington was appointed commander of the 74-gun ship "Orion." Participated in the Battle of Trafalgar and was awarded a gold medal. In 1808, Cordington commanded the Blake, and the following year he was assigned to an expedition against Walcheren Island. In 1810–1811 Cordington participated in the defense of Cadiz and cruised along the coast of Spain. In 1812, Cordington took part in a land battle against the French at Ville-Sukna. In 1813 Cordington was appointed colonel of marines. He soon sailed to North America, flying the broad pennant on the 40-gun frigate "Fort." In 1814, Cordington was promoted to the rank of Rear Admiral of the White. Participated in the attack on Washington and New Orleans. In 1821, Cordington was promoted to the rank of vice admiral of the Blue. In 1827, Cordington, commanding the Mediterranean British naval forces, together with a French squadron, forced Ibrahim Pasha to rule out an armistice. When it was violated, Cordington, taking over the main command of the combined British, Russian and French fleets, and locked the Turkish fleet in Navarino Bay, where the battle took place, which ended with the destruction of the Turkish fleet. According to legend, King George IV, dissatisfied with the complete defeat of the Turkish-Egyptian fleet, which gave significant advantages to Russia and disrupted relations with Turkey, on the margins of the decree awarding Admiral Cordington with the Most Honourable Order of the Bath, wrote: "I am sending him a ribbon, although he deserves a rope." In 1830 Sir Cordington was promoted to the rank of vice admiral of the White, and in 1837 to the rank of admiral of the blue flag. In Greece in 1927 – the year of the centenary of the Battle of Navarino – a postage stamp was issued with his image (Fig. 7).

Fig. 7 British admiral Codrington Edward. (Source: https://thumbs.dreamstime.com/b/edward-codrington-26656236.jpg)

Colonne Capo

Colonne Capo – one of the toe caps of the "Apennine boot" of Italy. The height of the promontory is 20 m. It is rocky and is situated 9 km southeast of the port of Crotone. A lighthouse, a large quadrangular tower, several houses, and a chapel are installed on the cape. There are the ruins of an ancient temple with a large colonnade nearby.

Comber (*Serranus cabrilla*)

Comber (*Serranus cabrilla*) – Comber (Serranus cabrilla) – a sea fish, in the Serranidae family. Its head is covered with scales from above and from the sides, the mouth is rather large, the body is brown-yellowish or brown, on the sides with 5–8 darker transverse stripes. Its length is up to 40 cm. Weight is approximately 0.5 kg. Usually, the length of the fish reaches 20 cm in catches.

It reaches sexual maturity at the fourth year. It keeps mainly on stony and rocky soils with thickets of underwater vegetation. It is a predator. It feeds on relatively large fish, as well as crustaceans. The meat is very delicious and tender.

Common Bottlenose Dolphin (*Tursiops truncatus*)

Common Bottlenose Dolphin (*Tursiops truncatus*) – The most famous and most completely studied dolphin species. It is more correct to call the front part of the dolphin formed by the jaws and intended for capturing food not "nose" but "beak" or "rastrum," since the olfacting there is absent. A dolphin breathes through a hole (a skin-muscular valve), called the "blowhole," which is located in a small hole on the sinciput. Adult bottlenose dolphin usually has a length of 2–2.5 m and a weight of 130–200 kg; however, it can reach a length of 3.8 m and a weight of 650 kg (off the coast of Great Britain). Bottlenose dolphin is common in temperate and tropical latitudes around the world. This type of dolphin is considered coastal, although it is also found in open areas of the ocean. Life expectancy is on average about 30 years; however, it can reach 40 years. An adult bottle-nosed dolphin eats 8–15 kg of various species of fish, squids, and crustaceans per day. Dolphins are incredibly maneuverable, capable of performing precisely planned jumps into the air up to 5 m high and up to 10 m long, and also at a speed of up to 40 km/h, instantly change the direction of movement or stop completely at a braking distance of 1–2 m. They dive to a depth of 300 m, remaining under water up to 15–20 min. The dolphin is one of the most amazing and mysterious living beings on our planet, a unique creation of nature, ideally adapted to life in the marine environment, endowed with a highly developed mind and sensory system, in its parameters far ahead of existing technical systems. Throughout the history of mankind, people knew that dolphins help and rescue drowning people and disperse sharks, gathering near humans. They especially protect children. A unique specificity of dolphins is their goodwill, sociability, and desire to establish contact with humans and help them. They are intelligent, good-natured, and even somewhat condescending toward people. For 2500 years of communication with a dolphin, there has not been a single case of aggressiveness of the latter, even if the man would hurt them. Even the bottle-nosed dolphin has 88 sharp conic teeth like tiger teeth. It could easily, for example, bite a man's hand off or kill him with a strong blow of the lower jaw, saving its life (Fig. 8).

Common dentex (*Dentex dentex*)

Common Dentex (*Dentex dentex*) – toothed sea fish of the Sparidae family, a typical representative of white fish and one of the most important representatives of commercial and sport fishing. The body is oval, high, flattened from the sides. Length is up to 1 m, weight reaches 16 kg, and average weight is up to 1 kg. It is found throughout the Adriatic Sea on a rocky bottom to a depth of 200 m. Most often it occurs at depths from 5 to 30 m. Avoids turbid water and sandy bottom areas. It is harvested all year round; however, it is better to catch it in spring and summer. Catch in the Adriatic Sea may reach 50–60 t, and in the Mediterranean Sea approximately 6000 t. Meat is delicious and valuable (Fig. 9).

Common Smooth-Hound (*Mustelus mustelus*)

Common Smooth-Hound (*Mustelus mustelus*) – a widespread species of cartilaginous fishes of the family Triakidae. The maximum recorded length is 200 cm. They are dwelling at a depth of 5–50 m, although there are records of the presence of these sharks at a depth of 350 m. A study conducted in 2000 in the waters of the seas washing the shores of Italy showed that most of the common smoothhounds population is concentrated in the Adriatic Sea and near the southern coast of Sicily.

Fig. 8 https://web-zoopark.ru/wp-content/uploads/2018/07/4-53.jpg

Fig. 9 Common dentex. (Source: https://x-dive.ru/files/upload/2018-08-13_22:42:02_3c490ad750401ef8.png)

Common smooth-hound has a short head and an elongated body. The distance from the tip of the snout to the base of the pectoral fins is from 17% to 21% of the total body length. The snout is slightly elongated and blunt. Oval large eyes are stretched horizontally. At the corners of the mouth are labial grooves. The mouth is rather short, almost equal to the eye, and its length is 2.2–3.5% of the body length. Blunted and flat teeth are asymmetrical, with a small central point. The distance between the dorsal fins is 18–25% of the body length. The pectoral fins are large, the length of the front edge is 13–17%, and the posterior edge is 8.2–14% of the total length, respectively. The length of the anterior border of the ventral fins is 6.5–9.9% of the total body length. The first dorsal fin is larger than the second. Its base is located between the bases of the pectoral and pelvic fins. The tail fin extends almost horizontally. Its color is gray or graybrown without marks. The belly is bright. Reproduces placental live birth. In addition, embryos feed on yolk. There are from 4 to 15 newborns in brood. Males and females reach sexual maturity with a length of 70–74 cm and 80 cm which corresponds to the age of 9.1 and 10.7 years. Pregnancy lasts 10–11 months. The length of newborns is about 39 cm. The maximum recorded life expectancy is 25 years. The diet consists mainly of benthic crustaceans such as lobsters, crabs, and shrimps. In addition, common smooth-hounds eat cephalopods and small bony fishes. The species is not dangerous to humans; however, you should be careful with these sharks, when you catch them, as they have sharp teeth, powerful jaws, and very rough skin. Meat is used for food in salted or fresh form, the liver is used to produce fat, and the skeleton is processed into fish flour. The World Conservation Union has given this species "Vulnerable" status.

Common Thresher, or Fox Shark, Zorro Thresher Shark (Common Thresher, *Alopias vulpinus*)

Common Thresher, or Fox Shark, Zorro Thresher Shark (Common Thresher, *Alopias vulpinus*) – the largest species of cartilaginous fish of the fox shark genus of the cognominal (Alopiidae) family. It reaches 7.6 m. Half of this length contains the elongated upper lobe of the caudal fin. These sharks have a streamlined body, a short and pointed snout, and medium-sized eyes. Fox sharks live in all temperate and tropical waters, although they prefer a cool temperature. They are found both in the open ocean at a depth of up to 350 m and near the coast, and usually keep in the surface layers of water. Fox sharks make seasonal migrations and spend the summer at lower latitudes. Weight is up to 400–500 kg. The teeth are small. The eyes are small. The ration mainly consists of schooling pelagic fish such as bluefish, mackerel, herrings, sargans, and luminous anchovies.

Fox sharks hunt using their long tail as a whip. They knock, fold, and stun their prey. This explains their name in English, "thresher shark." These are powerful and fast predators capable of jumping completely out of the water. Their circulatory system is modified to maintain metabolic heat energy and heat the body above the temperature of the surrounding water. Breeding takes place through a placental live birth. Mating occurs in the summer, usually in July and August, and childbirth proceeds from March to June. Pregnancy lasts 9 months. Fertilization and embryo development occurs in utero. There are up to four newborns in litter. The length of the newborn is 114–160 cm and is directly dependent on the size of the mother. Young sharks add 50 cm per year, while adults grow only 10 cm. Life expectancy is at least 15 years, and the maximum period is about 45–50 years. Despite its size, it is believed that fox sharks are not dangerous to humans, as they have small teeth and timid disposition. This species is an object of commercial fishing and sport fishing. Their meat and fins are highly valued. The low level of reproduction makes common sea foxes very sensitive to overfishing. Meat, especially fins are highly valued. It is consumed fresh, dried, salted, and smoked. The skin is excreted, and vitamins are produced from the liver fat.

Corfu

Corfu – The Italian name of the Greek Kerkyra Island.

Corinth Canal

Corinth Canal (*Korinthou canal*) – connects the Saronic Gulf of the Aegean and the Gulf of Corinth of the Ionian Seas. It was named after the city of Corinth, located at the western tip of the canal. The length of the canal is 6 km, and the depth is 8 m. The walls of the canal are of natural origin, made of limestone, and the height of the walls reaches 76 m. The width of the canal at sea level is 25 m, and at the bottom of the sea – 21 m. One railway and three road bridges are overthrown above the canal (Fig. 10).

In addition, submersible bridges function on both sides of the canal. Since ancient times, attempts have been made to build a canal in this place. The first mention dates back to the seventh century BC, when the Corinthian tyrant Periander, included among the seven wise men of ancient Greece, tried to dig a waterway, but ceased the work. The reason for the cessation of work was Periander's fears that due to the discrepancy between the water levels of the Aegean and Ionian Seas, flooding of the lands is possible. According to another version, he was afraid to arouse the wrath of the gods, because the oracle in Delphi proclaimed: "Do not supply the isthmus with a tower and do not break it (do not cross it with a channel)." Some assume that this divination came from the priests of the Corinthian temples, who feared that due to the opening of the channel for the rapid passage of ships, they would lose generous offerings and gifts, because the merchants would no longer have a reason to stay in Corinth.

Fig. 10 The Corinth Canal. (Source: https://cdn-dvizhenie24.jq7.ru/2020/02/kanal.jpg)

Instead of a canal, the tyrant created a simpler and less expensive stone portage, called "Diolkos." The remains of Diolkos currently exist next to the canal. Demetrius Poliorketus thought about the possibility of laying a canal in 307 BC. However, he also abandoned this thought when the Egyptian engineers, whom he hired to carry out his plans, assured him that the water level in the Gulf of Corinth and in the Saronic Gulf was not equal, and if a strip of land was dug, the waters of the Corinthian Gulf would rush into the Saronic Gulf and flood the entire area and flush the neighboring islands. When Corinth came under the rule of the Romans, Julius Caesar and later Caligula also developed such plans. In 67 AD Emperor Nero made a third attempt to dig a canal, employing 6000 slaves and convicts, and the emperor personally participated in the beginning of construction, being the first to throw away a clod of earth with a shovel. The construction of the canal became one of the reasons for undermining the economy of the state, depleting the provinces, which led to an uprising, the outcome of which was the death of Nero. After his death, the expensive project was abandoned. In subsequent years, Herodes Atticus and later the Byzantines tried to cross the isthmus, but their efforts were unsuccessful. The same can be said for the Venetians, who started work but soon gave up again. After the Greek Revolution in 1821, the first president of Greece, Ioannis Kapodistrias, also makes plans for the construction of the Corinth Canal. After the opening of the Suez Canal, the Greek government in November 1869 issued a law "on crossing the Isthmus of Corinth by the canal." The design of the canal was entrusted to the Hungarian architects Istvan Thürr and Bel Gerster, who had previously prepared the project for the Panama Canal. On May 5, 1882, work on the construction of the canal began. The construction of the canal was started by a French company, which stopped working due to financial difficulties, and never finished it. A Greek company led by Greek banker and philanthropist Andreas Syngros took over the project and completed the project in record time. The operation of the canal began in 1893. The canal allowed ships that circled the Peloponnese to shorten their journey by more than 400 km. The construction was carried out for over ten years (1881–1893) by the forces of 2500 workers. On August 7, 1893, a celebration was held to mark its opening. During the German Nazi invasion of Greece on April 26, 1941, they landed paratroopers on both sides of the canal, which, catching the few Allied forces defending it by surprise, captured and cleared the only bridge at that time. However, the explosive charges themselves were not removed in case of a need to retreat, and the surviving defenders, firing a cannon at the bridge, hit one of the charges, as a result of which the bridge was blown sky-high. It took German combat engineers two days to build an accommodation bridge over which, on 28 April, tanks of the German fifth armored division passed. The seizure of the isthmus led to the capture of 12,000 Allied soldiers who remained on the mainland, but due to the delay in the landing and the explosion of the bridge, the British Mediterranean fleet managed to evacuate 50,000 people from the Peloponnese. During the evacuation of troops from Greece in October 1944, the Germans filled it up with more than 645,000 m^3 of soil by explosions, debris from a railway bridge, and the remains of abandoned equipment, including rolling stock, mines, and the steam ship "Vesta," which was blown up and sunk in the canal. In addition, in the absence of maintenance, the water area was brought in by sediments significantly. American enterprises were involved in the restoration work. Clearing began in November 1947. After 8 months of operation, on July 7, 1948, the canal was opened to shallow draft ships, and in September to all types of ships. Currently, the channel has lost some of its economic importance. Due to the narrowness of the waterway, reverse movement is organized in it. Large ocean-going vessels with a width approaching 20 m cannot pass the channel. Large-tonnage vessels pass the channel in tow, as there is a danger of erosion of the walls. Annually, 11,000 ships pass through the canal. Bungee jumping is practiced on the bridges of the canal.

Corinthian Bay

Corinthian Bay (*Korinthou Limani*) – the southeastern part of the Gulf of Corinth. It forms an inlet of 15 km into the southern part of the northwestern coast of the Isthmus of Corinth. The greatest width of the bay in its middle part does not exceed 9.5 km. The northern coast is the slopes of the peak of Melangavi. At the top of the bay begins the low-lying coast of the Isthmus of Corinth. The southern and southwestern shores are lowland. The depths in the middle part of the bay exceed 90 m. At the top of the bay, 3.5 km southwest of the western entrance to the Corinth Canal, there is a large port city of Corinth. The entire coast of the bay is occupied by small settlements.

Cosmas of Aetolia (1714–1779)

Cosmas of Aetolia (1714–1779) – secular name Kosmas (*Cosma Etolicfs*) (1714–1779) – priest of the Constantinople Orthodox Church, a figure of Greek enlightenment of the eighteenth century. Outstanding Representative of Aetolia. He was born near Thermo in the Peloponnese and received the name Konstas. His surname is not known. Cosma's parents were weavers. He spent his childhood in poverty. Until the age of 20 he did not know the alphabet. Thirsty for knowledge, Cosmas turned first to the local priest Ananias. Later, having moved to the village of Agia Paraskevi, he continued his studies with the teacher Theophanes. Here he lived for 8 years, and became Theophanes's assistant and teacher in the village of Lobotina. In 1743 he went to Athos to the Athos Academy, after which he retired to the Philotheou monastery, where he took the monastic vows with the name of Cosmas, and later was ordained a hieromonk. He spent two years in the monastery. In 1759 he left the monastery and went to Constantinople, where the Patriarch of Constantinople Seraphim II (who occupied the throne in 1757–1761) gave permission to Cosmas to preach the Word of God. Thus, Cosmas began his enlightenment path, which made him famous and led to his martyrdom. In his sermons, he encouraged parents to teach their children Greek, which was the "language of the church." Taking upon himself the apostolic mission, he addressed the Orthodox in the spoken language of that era that was accessible to them. For 16 years, having passed through Thrace, Macedonia, Thessaly, Central Greece, the islands of the Aegean archipelago, and the Ionian Islands controlled by the Venetians, and by the end of his activity concentrated his efforts on Epirus and Albania, he created more than 100 schools. Already during his lifetime, Cosmas was called among the people the new apostle. Cosmas's sermons resonated not only among the Orthodox population, but also among the Muslims. At the same time, Cosmas aroused hostility among the Greek nobility, higher clergy, and Jewish merchants, as he undermined their economic and social foundations. After the defeat of the Ottoman Empire that began due to the actions of the Russian squadron (First Archipelago Expedition), the uprising in 1770 in the Peloponnese and in other Greek regions broke out. Cosmas was among those who came to the conclusion that the Greek people should no longer rely on outside help, but should focus on their own strength. After arriving in Berat (now southern Albania), Kosma was charged as a Russian agent. He was arrested and hanged in 1779. It is believed that the charge, arrest, and execution were provoked by Jewish merchants who suffered losses from the activities of Cosmas. In 1813 Ali Pasha of Tepelena, ruler of Epirus, ordered to build a church on the place of the execution of Cosmas in Kolkondas near the mouth of the Seman river, where his remains were buried. Cosmas was canonized by the Ecumenical Patriarchate on April 20, 1961, and on August 24, the day of his martyrdom is celebrated in his memory. In 1984, the remains of Cosmas were transferred to the archaeological museum of the Albanian city of Fier. Other relics of the saint were transferred to the Metropolitan Cathedral of Athens. For Greek historians, the main contribution of Cosmas is the enlightenment of the people in the era of a foreign yoke. For the church, Cosmas is a saint and a martyr who contributed to the preservation of Orthodoxy. Those, who on one way or another, related to the Church emphasize his miraculous abilities and prophecies. Some of the prophecies of Cosmas are heard at the beginning of the twenty-first century surprisingly actual. Thus he said: "The whole world will be girded with one thread, people will talk from one distant place to another, for example, from Constantinople with Russia"; "Thieves and robbers will live in cities, dress like ordinary people, and come in broad daylight to rob you"; "The time will come when demons will penetrate into dwellings under the guise of small boxes, and their horns will stick out on the roofs." In the context of the debt crisis that began in 2010 in Greece, the following prophecies of Cosmas are perceived in a completely different way: "They will lend you a lot of money and demand it back, but cannot take it"; "You will be imposed a large, unbearable tax, but they will not be able to get their way." Cosmas also told about the iron birds that sometimes peacefully fly in the sky, and sometimes they spew fire; about several villages that are located in one house; about the devil who hid in the box and shouted from there; about the abundance of inedible food (we are talking about TVs, telephones, military aircraft, etc.) (Fig. 11).

Fig. 11 Priest Cosmas of Aetolia. (Source: https://politobzor.net/uploads/posts/2018-02/1519674692_grecheskiy-svyatoy.jpg)

Crabs *Brachyura*

Crabs, *Brachyura* – suborder of invertebrates of the decapod crustacean order. The head is small with eyestalks. Cephalothorax is broad, and the width of chest shield varies from 2 to 20 cm. The first pair of ambulatory legs is equipped with claws. The abdomen is short, tucked under the cephalothorax; abdominal limbs of males (two pairs) are transformed into a copulative apparatus; of females (4 pairs), they are used for carrying eggs. They dwell in the seas, in freshwater, and on the shore. All crabs, except freshwater, breed in the sea. Crab's development occurs with metamorphosis; the larva zoea comes out of the eggs, which turns into a larva megalopa, and then into an adult Crab. Only freshwater crab (family Potamidae) does not have a free-floating larva. Pursued crab is able to break off limbs with a sharp movement, and the new limbs grow instead of the old ones. Crabs feed mainly on invertebrates. Many crabs are edible and serve as an object of fishing. Some crabs eat commercial mollusks – oysters and mussels. There are more than 4000 known species of crabs. Crabs contain more than 20% of the marine crustaceans that are caught, grown, and sold worldwide – approximately 1.5 million tons/year. Crab meat is specified by a low-calorie content, since 100 g of such meat contains 73 kcal. There are 96 kcal per 100 g in the boiled crabs, and the energy value of canned crabs is 85 kcal to 100 g. In 100 g of steamed crabs, there are 97 kcal. Crab meat has a large amount of proteins and nourishes the body with energy well. In moderate amounts, this product does not cause adiposis. Row crab meat has a jelly-like consistency and a grayish color. Edible crabmeat is in the limbs and abdomen. The limbs consist of armored tubes connected by joints. Under the shell, crab's body is covered with a red film. It has been the basis for the forming of a new shell after crab molt, if it would remain at sea. Crab meat is not only dietary but also rich in vitamins, amino acids, and micro-elements. These include iodine, calcium, zinc, polyunsaturated fatty acids, and antioxidants: taurine and vitamins E and C. Also, crabs are useful in anemia, cardiovascular diseases, and in visual deterioration. A number of high-quality products may be produced from crabs: canned goods, boiled-frozen crab legs, dried meat. Realization of fresh and cooled crabs must be made very quickly: at a temperature not higher than 12–15 °C, crabs are stored for no more than 15 h, and iced they are preserved for 30–36 h. (Creek caravides.) (Fig. 12).

Fig. 12 Crabs, *Brachyura*. (Source: https://upload.wikimedia.org/wikipedia/commons/thumb/0/06/Jonah_crab_%2811823580556%29.jpg/1200px-Jonah_crab_%2811823580556%29.jpg)

Cretan Sea

Cretan Sea (*Kritiko pelagos*) – part of the Mediterranean Sea between Crete and the Cyclades. This is the smallest sea of all that wash the shores of Greece. In the northern part, it borders the Aegean Sea. In the west and north, through the straits of Antikythera and Kythira, it borders the Ionian Sea. In addition to Crete, there are several other large islands in the Cretan Sea: Kasos, Anafi, Santorini. The Cretan Sea is very warm, in August the water temperature rises to 26°. It cools down slowly, allowing the holiday season to last until the end of October, starting around mid-April or early May. In winter, the sea temperature is around 15°. The coast of the Cretan Sea is gently sloping, smoothly descending into the water, forming shallow beaches. The Cretan Sea is loved by divers for its diversity of fauna and perfect water purity. Among the representatives of the animal world there are crabs, lobsters, sea urchins, sponges, tropical fish, and other inhabitants. The Cretan Sea is one of the most popular for beach recreation in Greece (Fig. 13).

Crotone

Crotone – city located at the western shore of the Taranto Gulf of the Ionian Sea in Calabria, Italy. Founded by the Greeks in 710 BC. According to legend, the city was founded on the advice of the Delphic Oracle. The ancient Croton got its name in honor of the son of the king of the island of Aegina Eak – Croton. The ancient settlement occupied a hill, where, over time, the acropolis, traditional for the Hellenic policies, arose. Since 580 BC, krotonians shone at the Olympic Games (Milo of Croton). In 530 BC Pythagoras founded a philosophical and ethical school in Kroton, which supported the local oligarchic government. In 510 BC Croton defeated its main rival – the legendary Greek city of Sybaris. The tyranny of Cleinias was established shortly thereafter. The Pythagoreans left Croton and tyranny was replaced by people's rule. At the end of the sixth century BC Croton for a short time became the most powerful city of Magna Graecia. In 379 BC Croton for 12 years came under the rule of the Syracuse tyrant Dionysius I. In 295 BC in Croton, the tyrant Menedemos ruled, who was expelled by the Syracuse king Agathocles, after which the city briefly became part of his Syracuse state. In 277 BC Croton was conquered by the Romans, and in 194 BC it became a Roman colony. After the fall of the Roman Empire, the city became Byzantine. In 870 it was plundered by the Saracens. In the sixteenth century the Spanish king Charles V granted the city special privileges and built a port. In the sixteenth century the city began to be called Kroton. And only in 1928 it changed its name to the modern one – Crotone. In times of Mussolini, the city became the largest industrial center in Calabria, favored by the proximity of a hydroelectric power plant. In the 1980s. due to the bankruptcy of the main factories, it experienced

Fig. 13 Cretan Sea (*Kritiko pelagos*). (Source: https://i.pinimg.com/originals/c4/77/70/c4777039a2cb32827051dfd33cf99404.png)

Fig. 14 Crotone. (Source: http://www.lasty.com.pl/city/large/0/LECI/le-castella.JPG)

an economic crisis. In 1996 it suffered from flooding. Since 1955 archaeological excavations have been conducted in Crotone. The port of Crotone is found at a cape of 43 m high located 12.5 km in the south of the mouth of the Neto River, which flows into the Ionian Sea. The port consists of two harbors: Porto Nuovo and Porto Vecchio. Crotone is a city of modern buildings. The castle of Charles V, which is now a city museum, where, among other things, the results of the latest archaeological excavations are exhibited, a small fortification of La Castella on an island in the bay and the Romanesque Cathedral remind of the ancientry of the city (Fig. 14).

Cyclops Riviera

Cyclops Riviera – the coast of the Ionian Sea near the town of Aci Castello and Aci Trezza. The name is taken from the legends about the ancient Greek hero and traveler Odysseus (or Ulysses). Cyclops lived in Aci Castello many thousands of years ago, one of them – Polyphemus – was the beloved son of Poseidon. When the fearless Odysseus, saving himself and his friends, blinded the offspring of the sea god, the enraged Cyclops chased Ulysses and began throwing gigantic stones at him. These four giant basalt rocks remained where Polyphemus threw them – in the Gulf of Catania, north of Catania, opposite the village of Aci Trezza, 2–3 km away. They have the shape of a prism of a dark color, but are clearly visible against the background of a bright coast. The largest and northernmost of the rocks is the Aci rock with a height of 125 m. In the 1960s–1980s, there was a physical and biological station on it. There was even a castle on one of them in the eleventh century. Only ruins of this castle and one of the towers of the fortress have survived to this day. The passage between the rocks and the coast is shallow.

D

Daskalio

Daskalio – (1) an island located between Ithaca and Kefalonia islands, Greece. This is the only piece of reddish rocks in the strait. The height of the island does not exceed 3 m. There are ruins of a watchtower built in the Middle Ages on the island.

2) A small island located 1 km west of Ambelos island, Gulf of Corinth, Greece. The shores are composed of rocky cliffs. Its altitude is 26 m.

Democedes

Democedes (*Demokedes*) – Greek physician from Croton, Southern Italy. In the second half of the sixth century BC, he was one of the greatest representatives of Western Greek medicine. Herodotus mentions him as a doctor who surpassed all his contemporaries. He was a doctor in Aegina and Athens, and later a doctor of the tyrant Polycrates of Samos and the Persian king Darius. Then Democedes returned to Croton and, possibly, joined the Pythagoreans.

Diolkos

Diolkos (Diolk, diolkos) – a paved trackway built to shorten the sea route around the Peloponnese (352 km), Greece. Its length is 6 km, and it ran along the future route of the Corinth Canal. One of the most daring technical projects of Antiquity, an achievement that largely contributes to the economic development of the state. This grandiose structure was put into operation at the end of the seventh century or at the beginning of the sixth century BC during the reign of the Corinthian tyrant Periander. Diolkos, a road paved with stone slabs 6–7 km long, was made specifically for dragging ships across the isthmus from the Corinthian Gulf of the Ionian Sea to the Saronic Gulf of the Aegean Sea. The ships were dragged ashore, hoisted on wooden cylinders and, thus moving, lifted onto special carts (olks), which moved along parallel rails (furrows). After moving from one end of Diolkos to the other one, the ships were floated of the water in the same way. Moving was not easy, hence, to make the ship lighter, its cargo was unloaded, transported separately, and then reloaded at the other end of Diolkos. In the course of archaeological research of the Archaeological Service under the leadership of N. Verdeli in 1952–1956 fragments of Diolkos were discovered. In the southwestern part of the modern canal, the beginning of Diolkos, 10 m wide, in the Gulf of Corinth has been preserved. It can be seen in the water, next to the ancient pier. Then it rose toward the east, and was found at the southern edge of the channel. Parallel furrows for wheels (olki) are visible at a distance of 1.5 m from each other and are marked on a stone road 3.5 m wide. Perpendicular cuts are also observed – traces of ropes with which

I. S. Zonn, *The Ionian Sea Encyclopedia*, Encyclopedia of Seas,
https://doi.org/10.1007/978-3-031-08206-1

the carts were dragged. Diolkos route was not straightforward, as follows from another fragment found at the northern edge of the canal in one military unit. It went on a flat surface to avoid steep slopes. This technical achievement of the Corinthians, despite the fact that it brought significant revenues to the policy, was complex, expensive, and could not serve large ships. Therefore, Diolkos was used mainly for transporting warships during the crisis in the fifth century BC, during the Macedonian period and in the Byzantine era. Diolkos was used until the twelfth century.

Dionysios Solomos

Dionysios Solomos (*Solomos Dionisios*) (1798–1857) – famous Greek poet. Born on the island of Zakynthos, one of the Ionian Islands. In 1808 Solomos went to study in Italy. This was the period of the establishment of romantic tendencies in Italian literature, with their spirit of patriotism and love of liberty. Graduated from the Pavia University's Faculty of Law. Solomos wrote his first poems in Italian; however, having returned to Greece in 1818, he studied his native language, Greek folk songs. He creates his Greek poems on "Dimotiki" Greek language of the people. In 1821 he joined the ranks of the patriotic society Filiki Eteria. The first major work is Solomos's poem in Greek "Hýmnos is tin Eleftherían" – "Hymn to Liberty" or "Hymn to Freedom" (1823) (Fig. 1).

Fig. 1 Dionysios Solomos. (Source: https://libryansk.ru/files/media/2020/04/img_solomos.jpg)

The main works of the poet: the poems "Ode to the death of Lord Byron" (1825) and "The Free Besieged" (on which the poet worked until the last days), dedicated to the theme of the people's struggle. Although later works ("The Greek little boat," "Sappho," etc.) were written by Solomos in Italian, they also used motives of Greek folklore. The first stanzas of the "Hymn to Freedom" from 1869 became the words of the national anthem of Greece. Solomos died in 1857 in Kerkyra. In 1865 his remains were transported to the island of Zakynthos. There is also a monument to him and a museum.

Dolphinfish, *Coryphaena hippurus*

Dolphinfish, *Coryphaena hippurus* – genus of ray-finned fishes from the monotypic family of dolphinfishes (Coryphaenidae). Fish live in the Mediterranean, in the Atlantic, Pacific, and Indian Ocean. The body length of dolphinfish varies from 1.0 to 1.5 m. The body is elongated, compressed on the sides, covered with small roundshaped scales of silverish golden color. The maximum length of a large dolphinfishes is 2.10 m, with a maximum weight of 40 kg. The skull is blunt and rounded.

Dorado (Gilt-head bream) (*Sparus aurata*)

Dorada (Gilt-Head Bream) (*Sparus aurata*) – the fish of Sparus genus. Its length is up to 70 cm. Maximum weight is up to 8 kg. The body is oval, laterally flattened with a steep profile of the head. There is a distinct protrusion above the eyes. The fish has large serrated scales. Its head also covered with scales. Operculum has no spines, and the rim of the preoperculum is smooth. The inferior mouth is small. The color is ash-silver with a golden hue, which quickly turns pale after death. It is found in the Adriatic and the Mediterranean seas. From the time of ancient Rome to the nineteenth century it is an aquaculture object. The fish migrates over long distances. The meat is of average quality. It has a significant content of proteins, which makes it quite nutritious. Dorada is known

Fig. 2 Dorado. (Source: https://adriaticnature.ru/wp-content/uploads/2019/08/Sparus-aurata-1140x789.jpg)

to the world for a long time, and has always been the most popular fish of the Mediterranean. In ancient Rome, the meat of this fish was very much appreciated, so it was specially kept and fed in salt ponds and pools. In Venice, this method of fattening existed at the end of the nineteenth century. Artificial raising of this delicious fish has been developed since very ancient times and is now widely used in France, Italy, Turkey, Greece, and other countries. To obtain tender meat and excellent taste, it is bred in exceptional conditions: in closed rooms with a special light that imitates a certain time of the year. Catch in the Mediterranean is approximately 800 t (Fig. 2).

Dorians

Dorians (*Doris*) – one of the four main ancient Greek tribes. According to generally accepted views, they moved to Greece during the "Dorian invasion." In historical time, they were engaged in agriculture and cattle breeding, settled in Central Greece, in the Peloponnese (Laconia and Messinia), in southern Italy and in Sicily. In the Peloponnese, the Dorians became the dominant people.

Dörpfeld Wilhelm

Fig. 3 Dörpfeld Wilhelm. (Source: https://everipedia-storage.s3.amazonaws.com/ProfilePicture/en/Wilhelm_D_u_25C3_u_25B6rpfeld__5abb8a/Doerpfeld2.jpg__38440.jpeg)

Dörpfeld Wilhelm (*1853–1940*) – famous German architect and archaeologist who studied architecture at the Berlin Academy of Architecture. He took part in numerous excavations, including in Olympia, Troy, Mycenae, Tiryns, Athens, Pylos, and Corinth. While reading Homer's "Odyssey," Dörpfeld doubted the localization of Odyssey's homeland – the island of Ithaca. In his opinion, this should be the island of Lefkas (Lefkada), to which he dedicated his work "Alt Ithaka" (1927). In this work he reported on his excavations in Lefkada which lasted for 10 years (Fig. 3).

The founder of modern studies of buildings in classical archeology. Being a close collaborator of Schliemann (who discovered Troy), he made a great contribution to the development of scientific methods of excavation. Buried on the island of Lefkada, Greece. Dörpfeld's main works are "Troy and Ilion" (1902), "Ancient Olympia" (1935), and "Ancient Athens and Their Agora" (1937–1939). A monument was erected in his honor on the embankment in Nydri, and one of the streets is named after him.

Doukato

Doukato – the cape, the southwestern tip of the island of Lefkada, in ancient times was called the Lefkadian rock or Cape Lefkada, Greece. Steep, tall, and devoid of vegetation according to local legend, in this place, the famous lyrical poet Sappho committed suicide by jumping into the sea because of unanswered love. Not far from the cape there are the remains of the ancient temple of Apollo Lefkatas. A lighthouse is installed here. The church of St. Nicholas was built on the place of an ancient temple in the seventeenth century.

Drakonera

Drakonera – the island is the largest of the Dragonera group of islands, Greece. Length from north-west to south-east 2200 m, width is 1500 m, and height in the middle part 130 m. The rocky slopes of the hill come close to the eastern coast of the island and descend steeply to the water.

Drakoneras

Drakoneras (Dhragonares, Drakonera Islands) – they are situated at 5.5 km north of the Echinades in front of the entrance to the Astakos Bay of the Ionian Sea on the mainland of Aetolocarnania, Greece. The islands are mostly high. Made up of rocks and boulders. There are navigable passages between them. Covered with maquis and olive trees. Most of them are cultivated. Drakoneras can be divided into two groups: southern and northern. The southern group includes Stamodi and the islands of Apasa, Soros, and Gravaris, located 4.5 km west of the Petalas Island. The northern group consists of a number of small and large islands and rocks. The most significant islands are Pontikos, Provati, Karlonisi, Kalogiros, and Drakonera.

Durrell Gerald Malcolm

Durrell Gerald Malcolm (1925–1995) – English naturalist, writer, and founder of the Jersey Zoo and the Wildlife Conservation Fund, which now bears his name. He was the fourth and youngest child of the British civil engineer Lawrence Samuel Durrell and his wife Louise Florence Durrell. According to the testimony of relatives, Gerald has fallen ill with "zoomania" already at the age of two, and his mother remembered that one of his first words was "zoo." In 1928, after the death of his father, the family moved to England, and 7 years later, on the advice of Gerald's older brother, Lawrence, to the Greek Corfu Island. The home teachers included the naturalist Theodore Stephanides (1896–1983) (Fig. 4).

It was him, who gave Durrell the first systematized knowledge of zoology. In 1939 (after the outbreak of World War II) he returned to England with his family and got a job at the London store "Aquarium." Durrell's career as a researcher began with work at Whipsnade Zoo in Bedfordshire. Here Durrell got a job immediately after the war as a "student keeper." Here he received his first professional training and began to collect a "dossier" containing information about rare and endangered species of animals (20 years before the appearance of the International Red Book). After the end of the war, he decides to return to

Fig. 4 Durrell Gerald Malcolm. (Source: https://api.rastimdetey.ru/storage/articles/363/icon_large.jpeg?1580455731)

his historical homeland – to Jamshedpur, India. In 1947, Durrell, having reached the age of 21, received part of his father's inheritance. With this money, he organized three expeditions – two to British Cameroons (1947–1949) and one to British Guiana (1950). These expeditions did not bring him profit, and in the early 1950s Durrell finds himself without a livelihood and work. At this time, Durrell's older brother advises him to take up the pen, especially since "the British adore books about animals." Durrell's first story – "The Hunt for the Hairy Frogs" – had an unexpected success, he was even invited to personally read this work on the radio. His first book, *The Overloaded Ark* (1953), was about a trip to Cameroon and received rave reviews from both writers and critics. The author was noticed by major publishers, and the royalties for "The Overloaded Ark" and Durrell's second book, "Three Singles to Adventure" (1954) allowed him to organize an expedition to South America in 1954. However, a military coup d'etat took place in Paraguay at that time, and almost the entire collection of animals had to be left there. Durrell described his impressions of this trip in his next book – "The Drunken Forest" (1955). At the same time, at the invitation of his brother – Lawrence – Durrell was vacationing in Corfu. Familiar places evoked a lot of childhood memories – this is how the famous "Greek" trilogy appeared: "My Family and Other Animals" (1956), "Birds, Beasts, and Relatives" (1969), "The Garden of the Gods" (1978). Durrell has written over 30 books (almost all of which have been translated into dozens of languages) and directed 35 films. Thirty years later, Durrell managed to shoot in the Soviet Union, with the active participation and assistance of the Soviet side. The result was the 13-episode film "Durrell in Russia" (shown on the first television channel of the USSR in 1986–1988). In 1959, Durrell founded a zoo on the island of Jersey, and in 1963, the Jersey Wildlife Preservation Trust was organized on the basis of the zoo. Durrell's main idea was to breed rare and endangered animal species in a zoo with the aim of further settling them in their natural habitats. Nowadays, this idea has become a generally accepted scientific concept. If it were not for the Jersey Trust, many species of animals would have survived only in the form of stuffed animals in museums. Thanks to the Trust, the pink pigeon, the Mauritius kestrel, the golden lion tamarin and marmoset monkeys, the Australian corroboree frog, the radiated tortoise from Madagascar, and many other species were saved from total extinction.

E

Earthquakes in the Ionian Sea

Earthquakes in the Ionian Sea – Greece is one of the most seismically active countries on the European continent, situated in the complex border zone of the eastern Mediterranean Sea at the junction of the African and Eurasian tectonic plates, between which there is a small Aegean Sea Plate. The northern part of Greece is part of the Eurasian Plate, while the southern part is the Aegean Sea Plate. The Aegean plate is moving southwestward in relation to the Eurasian plate at a speed of 30 mm per year. The African Plate is slowly sinking under the Aegean Plate at a rate of 40 mm per year. In the north, at the junction of the Aegean and Eurasian plates, a divergent border forms, and in the south, a convergent border, which is referred to as the Aegean arc. It is here that strong earthquakes have always occurred in the entire history of observations. In the place where the African Plate diving under the Aegean Sea Plate, the epicenter of the earthquake is often located at great depths – more than 50 km, and the magnitude (m) of earthquakes is usually higher than 7 points. However, such earthquakes do not bring much destruction with them, although they are felt very well. Among the strong earthquakes in the Ionian Sea, one can note: in 1908 – Messina – the number of victims (h.v.) – 70,000–100,000; in 1953 – the area of the islands of Kefalonia and Zakynthos m. 7.2, h.v. – 476 people; in 1981 – Gulf of Corinth – m. 6.7, h.v. – 22 people; in 2006 – Kythira – m. 6.8; in 2008 – the Peloponnese peninsula – m. 6.5, h.v. – 2 people; in 2009 – the area of Ionian Sea 108 km south-west of Patras – m. 5.1; in 2013 – Kefalonia Island, Lixouri – m. 5.8; in 2015 – south-west of Lefkada island m. 5.2, h.v. – 3 people.

Eastern coast the of Sicily island

Eastern coast the of Sicily island – it stretches from the cape, which is the northeastern tip of Sicily Island, almost 200 km southeast to the Portopalo bay, Italy. The coast is mountainous from the Strait of Messina to the port of Catania. The ports of Messina, Catania, Augusta, and Syracuse are located on the Eastern Coast.

Eastern Mediterranean

Eastern Mediterranean – is situated between 18°–36° east longitude and 30°–40° north latitude. It includes the Ionian and Levantine basins, and the Adriatic and Aegean seas. Its volume is 7800 km^3. The Levantine Basin is the second largest in the Eastern Mediterranean. It is surrounded by Asia Minor, the northeastern African mainland, and the Cretan Archipelago. The Ionian Basin is connected with the Western Basin by the Sicilian Strait (1000 m), and with the Adriatic Sea – by the Otranto Strait. The Ionian

I. S. Zonn, *The Ionian Sea Encyclopedia*, Encyclopedia of Seas,
https://doi.org/10.1007/978-3-031-08206-1

Abyssal Plain (4000 m) is bounded in the west by the shallow Tunisian and Maltese plateaus, as well as the Calabrian and Sirtian Uplands to the north and south of it, respectively. The Greek (Hellenic) depression outlines the Cretan Archipelago, with its deepest areas up to 5000 m southwest from Greece and 4300 m in the Rhodes Basin, expanding to the east. The middle Mediterranean underwater mountain ridge stretches from the Ionian Basin to the Levantine Basin south of the Hellenic Basin. On a smaller scale, the bathymetric features of the Levantine Basin are represented by the Latakian (100–1500 m), Sicilian (1000 m), Antalyan (2000–3000 m), and Rhodes (4000 m) basins and the Anaximander (1500 m) and Eratosthenes (3000 m) seamounts. The slopes of the mountains almost approaching to the edge of the water, leaving a very narrow lowland strip. The width in the harbor area of Riposto (south of Taormine) reaches 3.5 km. Several rivers flow into the Ionian Sea, the largest of which are Pagliara, Savoca, and Atro. The western shore of the Gulf of Catania is flat, low, bordered by a sandy beach. The Simeto River flows from this bank into the gulf. To the north of the mouth of the river lies a fertile plain, and to the south there are hills and in some places occur wetlands. The southern shore of the gulf is high and steep. From the port of Syracuse to Cape Porto Palo, the coast in the northern part to town of Avola is rocky and steep, and further to the south in the Noto Gulf it is low. Almost all settlements in this area are situated away from the sea.

Echinades islands

Echinades islands – a group of 18 islands located 3.6–9 km west-northwest of the mouth of the largest river in Greece, the Achelous, which flows into the Ionian Sea. It includes three large islands - Vromonas, Nissos Makri (Makri), and Oxeia, as well as the islet of Kouneli or Makropoula. Echinades is associated not only with the history of Greece, but also with its mythology, being mentioned as 18 punished nymphs. According to Hesiod, they were turned into islands for not praising Achelous, the son of Tethys and Oceanus, who was considered the god of the river of the same name. In turn, they are historically associated with the naval battle of Nafpaktos (which is also known under the Venetian name of Lepanto), which took place in October 1571. After the victory of the Venetians in the Ottoman–Venetian War, the victors sent livestock breeders from the neighboring island of Ithaca to the islets, passing them into the hands of such powerful families like Petalas, Zavos, Karavias, and Pierros. The islands passed from hand to hand until they became the property of the Dendrinos and Grivas families. The last shepherd left the territory of Echinades in 1986. At present, the islands of Oxeia, Provati, Apasa, Girovaris, Soros, and Pistros were bought by the Emir of Qatar Sheikh Hamad bin Khalifa al-Thani. All islands are uninhabited. Echinades are declared a National Park of the Lagoons of Missolonghi-Aetolia and belong to the European ecological network Natura-2000, which prohibits the building of large structures here.

Elafonisos

Elafonisos – (1) "island of deers" - an uninhabited island located at the confluence of three seas: the Ionian, Aegean and Libyan, near the southern part of the Peloponnese peninsula. The area is 19 km^2. The population is 745 people. Strictly protected by the state. There are many lilies, junipers, ancient cedars, and centuries-old oaks growing here, it is forbidden to cut any plants, and you cannot take sand with you as a souvenir. It is forbidden to leave garbage here, to burn fires, there is no camping, and there are no hotels. You cannot stay on the island overnight, and the road to it is not asphalt, but unpaved. You will not find deer here, but in ancient times there were many of them, and this was a favorite hunting place (Fig. 1).

According to historians, Elafonisos was not always an island. It used to be a peninsula called "Onou Gnathos," which means donkey's jaw, probably because of its shape. Around the sixth

Fig. 1 Elafonisos. (Source: https://images.vanityfair.it/gallery/131419/Big/5c0faeb3-a2bb-4286-a07d-c3ac94785172.jpg)

century BC the region was hit by a massive earthquake. It was then that the ancient city of Pavlopetri went underground, and Elafonnisos became an island. Pavlopetri was an important seaport, and when sailing to the island by ferry, its remains can be seen underwater. Although Elafonisos is considered an archaeological site, tourists mostly come here for the amazing beaches. The most famous of them is Simos, which is located in the southern part of the island and is considered one of the most beautiful in the Mediterranean. The pristine atmosphere of calmness and silence is preserved here. Takes 11th place in the list of the most beautiful and cleanest places on the planet.

(2) The strait, which runs between the southern shores of the Elos peninsula and the island of Elafonisos, on the one hand, and the northeastern coast of the island of Kythera, on the other. The strait is used by ships sailing from the Aegean Sea to the Ionian Sea. The smallest width of the strait is 8.3 km. The strait is deep. There are no dangers, except a rock with a depth of 5.4 m, which lies near Cape Spathi.

Epirus

Epirus – nomos in Greece, located in the north of the Balkan Peninsula between the Adriatic Sea and Lake Ohrid from Scodra to Ambracian Gulf (Albania and Greece).

Ereikoussa

Ereikoussa – the northern island among the three large islands and it is situated to the northwest of the island of Kerkyra (q.v.). It is located 10 km east of the island of Othonoi (Fanò). The area is

4.5 km^2. The population is 698 (2001). The island is high. The height of its northern part reaches 121 m. The northern and western shores are cliffed, whereas the southern and eastern ones are sloping. A small bay with sandy shores forms an inlet into the southern coast. The shores are edged by a shoal cluttered with rocks.

Etesian

Etesian – steady wind from northern directions. Usually observed from about mid-May to mid-September in the area of the western coast of Greece. Its highest frequency is in July. The speed is usually 7–10 m/s, sometimes 15–20 m/s or more. Accompanied by dry and clear weather.

Europe

Europe – part of the world, which forms the continent of Eurasia along with Asia; it is its western part. In the north of it is washed by the seas of the Arctic Ocean – White, Barents, and Kara seas; in the west and south is washed by the Baltic, North, Norwegian, Mediterranean, Black, Azov, Caspian seas. The area of Europe is about ten million km^2, of which – archipelagos and islands with a total area occupy about 730,000 km^2, the largest ones: British, Iceland, Novaya Zemlya, Spitsbergen. The coastline is highly dissected, its length is approximately 38,000 km. About 25% of the territory of Europe falls on the peninsulas, the largest are Kola, Scandinavian, Iberian, Apennine, and Balkan. There are many bays and gulfs. The population is 750.5 million people (2021). There are 43 states that are completely located on the territory of Europe, while another five are located partially (Fig. 2).

European Barracuda or Mediterranean Barracuda (*Sphyraena sphyraena*)

European Barracuda or Mediterranean Barracuda (*Sphyraena sphyraena*) – sea pelagic fish of the Sphyraenidae family. The body has elongated,

Fig. 2 Europe. (Source: https://blocs.xtec.cat/ceipascoangles/files/2020/03/europe-political-map.jpg)

Fig. 3 European barracuda. (Source: https://upload.wikimedia.org/wikipedia/commons/thumb/c/c5/Sphyraenidae_-_Sphyraena_barracuda_%28Great_barracuda%29.JPG/1200px-Sphyraenidae_-_Sphyraena_barracuda_%28Great_barracuda%29.JPG)

cylindrical shape, covered with small cycloid scales. The head is elongated. Its mouth is large. The lower jaw is extended forward. Teeth are large. Eyes are round and large. The tail fin is forked. The back is greenish-blue, with transverse dark stripes above the lateral line. The belly is silvery white. It can reach a length of 1.8 m and a weight of up to 8 kg, and the usual length is up to 1 m. It occurs at depths of up to 100 m. It feeds on fish and mollusks. Meat is delicious, especially when frying. Because of the shape of the body, it is often called "sea wurst or sausage." (Fig. 3).

European bass (*Dicentrarchus labrax*)

European Bass (*Dicentrarchus labrax*) – a sea predatory fish of the Moronidae family. The body is elongated. The scales are large. There are two fins on the back. On the frontal fin, there are sharp spikes. The color is silver, and the back is gray or olive. The length of the fish is up to 1 m. It reaches a weight of 13 kg. The average weight is 0.5 kg. It is widely spread all over the coast of the Adriatic Sea. It is more common at the confluence of the rivers and the sea. It does not go deeper than 100 m into the sea, and it is usually found at a depth of 20–30 m. It approaches the coast for spawning. In recent decades, catches have been decreased. It is grown as aquaculture, in closed natural and artificial ponds with sea salt water. Such business is popular mostly in France, Tunis, Turkey, and Morocco. The meat is delicious. The European bass is often called the European sea bass. Sometimes European bass is called sea pike perch, wolf of the sea, capemouth, etc. (Fig. 4).

Fig. 4 European Bass. (Source: https://media.voog.com/0000/0039/9849/photos/Huntahven_Dicentrarchus%20labrax.png)

E

European conger (*Conger conger*)

European conger (*Conger conger*) – a large serpent fish of Congridae family. The body has a serpentine form, and the head is flattened. Skin is mucous that does not allow to take it in hand. The color is grayish or blackish. It can reach a length of 3 m. Weight up to 70 kg, average weight in catches – 1 kg. It breeds in the eastern part of the Atlantic Ocean over depths of more than 3000 m. Spawning has also been detected in the Mediterranean Sea. Adult fish after spawning die. The larvae spread over considerable distances and undergo a complex transformation. It is predator and eats fish. The meat is delicious (Fig. 5).

European Eel (*Anguilla anguilla*)

European Eel (*Anguilla anguilla*) – a fish of Osteichthyes (bony fishes) taxonomic order, Anguillidae family. It dwells in fresh water, and dies in salt water. It is easily distinguished from all others by a cylindrical, strongly elongated, serpentine body, covered with scales so small that they are not visible. The color of the back is brown or dark green, the sides are yellow, and the belly is yellowish or white. The eyes are small. They have numerous teeth. European eel is voracious predator, eating small fish, frogs, and crayfish. In winter, it digs deep into sludge and hibernates: high-fat content (up to 25%) allows it to go without food for a long time. Life cycle of the European eel has been a riddle for a long time, in the resolution of which great merit belongs to the Danish ichthyologist J. Schmidt (1922). It was found that adult eels roll into the sea from European rivers and head to the Atlantic Ocean, where in the Sargasso Sea at depths from 1000 to 3000 m with a water temperature not lower than +7 °C and salinity of at least 35‰ they spawn. Adult fish then die, and the leptocephalus larvae hatch which are then caught by the warm currents of the Gulf Stream and drift to the shores of Europe for 2 years, turning into glass eels (6.5 cm long). In the fourth year of life, they enter the rivers, where they acquire pigmentation. In rivers European eel live up to 12 years, reaching a length of 1 m and 4 kg of weight. Usually the average weight in catches reaches 200 g. Valuable commercial fish. Meat of European eel is extremely fat and has excellent taste, especially in smoked form.

Fig. 5 European conger. (Source: https://adriaticnature.ru/wp-content/uploads/2015/12/Conger-Conger.jpg)

European Hake (*Merluccius merluccius*)

European Hake (*Merluccius merluccius*) – benthic fish of *Merluccius* genus, Gadidae family. The color is silver-gray, and the top of the head and the back are dark gray. It is widely spread on the continental shelf of the Atlantic and Pacific oceans at depths of 100–100 m, several species – in the Mediterranean Sea. These are mainly bottom-dwelling fish that rise to both intermediate and upper layers of water in pursuit of prey. The length of the body of the hake can vary from 30 cm to 1.5 m. The body is banded. The head is large, 1/3–1/4 of the body length. The snout is elongated and compressed, the tip is wide and rounded. The eyes are large. The mouth is terminal. The lower jaw is longer than the upper one; the antennae on the chin is absent. It also has vomer teeth. The first complete ray of the dorsal fin is flexible and elastic, with 7–9 rays in the pelvic fin. The teeth on both jaws are well developed, sharp, arranged in two uneven rows. The upper teeth are fixed. Palatine teeth are missing. The gill rakers are well developed. The fish has two separate dorsal fins: the first is short, high, triangular in shape, and the second is elongated and partially divided by a notch in the middle part. The anal fin is similar to the second dorsal fin (Fig. 6).

The pectoral fins are long, thin, and set high. Their relative length becomes smaller as the fish grows. The caudal fin is smaller than the head; the fork increases with age. The surface of the body and head, except the anterior tip of the snout, is covered with fine and thin cycloidal scales. The lateral line is almost straight and continuous

Fig. 6 Evropian Hake (Merluccius, Merluccius). (Source: https://animalreader.ru/wp-content/uploads/2015/02/merluza-animal-reader.ru_.jpg)

throughout, consisting of 101–171 scales. The number of vertebrae is 48–58. The color during life is usually silvery, the back is gray-steel, the sides are lighter, and the abdomen is silvery-white. Some species are colored black. Sometimes they swim in river estuaries. Juveniles feed mainly on plankton and small crustaceans, and adult hake become predators and hunt pelagic schooling fish, including herring, mackerel and sardine, and squid. Cannibalism is observed. Almost all hake during feeding make vertical migrations, rising from the bottom at night. Some species migrate seasonally: in spring and summer they move to higher latitudes, and in winter they go deeper. Males grow more slowly than females, they have a shorter lifespan. Sexual maturity is usually reached by 3–4 years. Spawning in most species is extended and occurs in the summer months. Eggs are pelagic. Females spawn up to 500,000 eggs. The larvae first dwell in the water column, and then descend into the bottom layers. The number of many species has significantly decreased due to overfishing. Hake comes to the market fresh and frozen, some of the catch goes to the production of fish-flour. The meat is tender, white; there are few bones; and the boiled fillet is easily separated from the skeleton. When fresh, it loses its taste and aroma rather quickly. With the help of quick freezing this problem is eliminated (in Greek it is bakalliaros).

European pilchard (*Sardina pilchardus*)

European pilchard (*Sardina pilchardus*) – marine schooling pelagic fish of the *Sardina* genus of the Clupeidae family. The mouth is small. Behind the operculum on the body of the fish there is a spot, and several dark spots behind it. The length does not exceed 21 cm. The average weight is 30 g, usually 30–33 g. It reaches sexual maturity in a second year of life with a length of 13 cm. Spawning lasts from September to June, near the coast at a depth of 20–25 m at a temperature of 13–15 °C. Its roe are pelagic. It feeds on crustaceans and phytoplankton. The most important fishing object (Fig. 7).

Fig. 7 European pilchard. (Source: https://i.ytimg.com/vi/xxEN0IAUpHg/maxresdefault.jpg)

European plaice (*Pleuronectes platessa*)

European plaice (*Pleuronectes platessa*) – ray-finned fish of the Pleuronectidae family. The body shape of adults is oval, the body is strongly compressed. The head is small, less than a quarter of the length of the entire fish. The small mouth is in front of the head, the upper jaw is just below the right eye. Both eyes are on the right side of the body. The lateral line runs just above the pectoral fin. The anal fin has 48–61 soft rays, the dorsal fin has 65–84 soft rays, the pectoral fins have 7–14 soft rays, and the pelvic fins have 6 soft rays. The ophthalmic side of the body of the European plaice is olive brown blotched with conspicuous, but irregularly distributed, orange or yellow spots. The blind side of the body is white. The skin is smooth, even, with small scales. The maximum length can be up to 1 m, the average – 50–60 cm. The highest recorded body weight is 7 kg. Life expectancy is up to 50 years. Marine demersal fish. It occurs on sandy and silty soils on the continental shelf, more often at shallow depths (10–50 m), trying to hide from enemies during the day, burying itself in sediment. Sometimes it can dive into deeper (up to 200 m) layers. Juveniles keep near the coast in shallow water. The fish can tolerate sharp fluctuations in water salinity and survive in slightly salted and even fresh water bodies. Valuable commercial fish. The main food is polychaetes, crustaceans, and bivalves. Young (1–2 years old) individuals feed on shrimps and other small crustaceans (Fig. 8).

Fig. 8 European plaice. (Source: https://activefisher.net/wp-content/uploads/8/c/b/8cb51ccd49212c5525dea7fca539f4f5.jpg)

European Union (EU)

European Union (EU) – integration association of 27 independent states (February 2020): Austria, Belgium, Bulgaria, Hungary, Germany, Greece, Denmark, Ireland, Spain, Italy, Cyprus, Latvia, Lithuania, Luxembourg, Malta, the Netherlands, Poland, Portugal, Romania, Slovakia, Slovenia, Finland, France, Croatia, Czech Republic, Sweden, and Estonia. In January 2020, the United Kingdom became the first member state to leave the EU. The Treaty on European Union (EU) was signed in 1992 in Maastricht (the Netherlands) by the heads of state and governments of the 12 member states of the European Economic Community (EEC). The EEC was formed by Belgium, the Netherlands, Luxembourg, Germany, France, and Italy in 1958 with the aim of creating a common market for goods, capital, and labor by abolishing customs duties and other restrictions on trade. Later, Great Britain, Denmark, Ireland, Spain, Portugal, and Greece joined the EEC. Later, another 16 countries joined the EU. The main objectives of the integration process are set out in the Maastricht and Amsterdam (1997) treaties. These treaties define the transition of countries from the common market to economic and monetary unions, joint foreign and security policies, coordination of domestic, social and legal policies, EU expansion as candidates meet its requirements. To join the EU, the candidate country must meet the Copenhagen criteria, which were adopted in 1993 and confirmed in 1995 at the EU meeting in Madrid. The criteria require that states maintain democratic principles, the principles of freedom and respect for human rights, as well as the principles of the rule of law. There should be a competitive market economy in the country. The EU bodies are: the European Council, the European Parliament (with monthly sessions in Brussels and Strasbourg), the EU Council (Council of Foreign Ministers or other sectoral ministers), and the Commission of the European Union is a supranational executive body that carries out day-to-day work on the implementation of EU policies, Human Rights, Chamber of Auditors. Since 2002, a single currency, the Euro, has been introduced in the field of cash circulation among 19 EU member states.

Eurotas

Eurotas – river in Greece, on which the city of Sparta stands. It flows into the Gulf of Laconicos near the Rock of the Ionian Sea. According to legend, the hero Eurotas – the son of Lelex and the naiads of Cleocharia, or the son of Myles, the grandson of Lelex – took water into the sea from the plain that waterlogged it. A river stream, called Eurotas, arose. The tributary is Oenus.

Evinos

Evinos – river in Greece. Length – 80 km. The catchment area is 1070 km^2. The source is located on the slope of Vardousia. Further, the river flows between Vardousia and Panaitoliko, along Aetolia to the southwest, between Lake Trichonis (Trichonida) and the Nafpaktia mountains and flows into the Gulf of Patras 10 km southeast of Missolonghi. According to legend, the river was named Evinos, because Evenus, the son of Oceanus and Tethys, the king of the Aetolians, drowned in it, pursuing Idas, who abducted his daughter Marpessa. A water reservoir of the same name was built on the river. The dam and the tunnel connect the Evinos reservoir and the Mornos reservoir at the expense of Evinos. The volume of the reservoir is 14 km^3.

F

Ferry routes in the Ionian Sea

Ferry routes in the Ionian Sea – there is no ferry service between the Aegean and Ionian Seas. All ferry services are clearly limited to the waters of their own sea. Some shipping companies operate voyages in both seas, but such routes are separated and do not intersect with each other. There are two main ports for traveling in the Ionian Sea, – Patras in the northwest of the Peloponnese (220 km from Athens) and Igoumenitsa in Western Greece near the Albanian border (500 km from Athens), the third most important port is the port of Kylini, located a little further than Patras on southwest (280 km from Athens) (Fig. 1).

From these ports you can get to the islands of Zakynthos (only Kylini), Kefalonia (Kylini and Patras), Ithaca (Patras), Kerkyra (Corfu), Patras, and Igoumenitsa, as well as to Italy (some voyages lie directly from Patras, but most of them go to Igoumenitsa and Corfu). International ferry service is carried out with four Italian ports: Brindisi, Bari, Ancona, and Venice. Also, the Ionian Islands are connected with each other and the mainland by small local ferries. There are routes: Zakynthos – Kefalonia, Kefalonia – Lefkada, Kefalonia – Ithaca – Astakos (a small port on the mainland in the Akarnania region), in Kefalonia through the Argostoli bay between the two cities of the island – Argostoli and Lixouri, Parga (mainland, slightly to the south from Igoumenitsa) – Paxi island, and also between Igoumenitsa and Kerkyra.

Besides the large ferries, there are two or three local small ferry companies. Ferries from the Egeo town to Agios Nikolaos go across the Gulf of Corinth – an alternative to the Rion-Antirion bridge. In the southern Peloponnese, through the efforts of a local company, ANEN Lines, ferries operate between Kalamata (Messinia region), Gythio (Laconia region), the islands of Kythira and Antikythera, and Western Crete. The routes pass at the junction of the Ionian and Aegean Seas. Also, other local companies operate local flights from Neapoli (Laconia) to the islands of Kythera and to the island of Elafonisi.

Finikounda

Finikounda (Phinikunda) – a fishing village located in the south of the Messenia of the Peloponnese Peninsula between the towns of Methoni and Koroni. There is a small bay with an excellent beach, at the east – a large coastal beach strip, which has gained popularity among windsurfing enthusiasts.

Fish Cuisine

Fish Cuisine – Greeks do not eat river fish at all, only sea fish is preferred. In taverns, visitors choose fish visually. In order to do so, they are

I. S. Zonn, *The Ionian Sea Encyclopedia*, Encyclopedia of Seas,
https://doi.org/10.1007/978-3-031-08206-1

Fig. 1 Ferry routes. (Source: https://www.ferriesedipsos.gr/gallery/boat-posidonas-2-large.jpg)

allowed to enter the kitchen. In some seafood restaurants, the client can choose the fish from huge aquariums, which will be immediately caught and cooked. If you make a choice from the presented menu, you may see the mark "k," "kat," or an asterisk. This mark means frozen imported fish which is considered to be of inferior quality than the own freshly caught. Fresh fish (the Greek one) is simply more expensive than imported. The main way to cook large fish is charcoal baking. After baking, the fish is freed from large bones and served under a special sauce – "Ladolemono" (olive oil with lemon) (Fig. 2).

Fig. 2 Fish Cuisine. (Source: https://i.artfile.ru/2560x1706_1301313_[www.ArtFile.ru].jpg)

Boiled fish is served under the same sauce – the broth from it is the basis for the Greek fish soup, kakavia. Vegetables used in cooking kakavia (potatoes, carrots, tomatoes, onions) are mashed, and anise vodka is added to the broth during cooking. Small fish are often fried, and red mullet (mullet) and sole are only fried. Despite its very small size, the red mullet is considered a delicacy: it is not for nothing that it has the second name "mullet (sultanka)" – it is believed that in ancient Turkey only sultans had the right to eat it, and in ancient Rome it was the main dish at the feast of the patricians. This small fish is up to 25 cm long, and it is one of the most expensive fish in the Mediterranean. Red mullet fried on charcoal or grilled is very delicious. The fish that the Greeks prefer to eat fried include cod (Bakaliaros), swordfish (xifias), smelt (aferinia), and little shark (glaros). A feature inherent in Greek fish cuisine is, first of all, the savoro sauce. Olive oil is used for this sauce – certainly the one on which the fish was fried. Tomatoes, garlic, rosemary, and spices are sautéed with oil. One of the Greek ways of baking fish is interesting, which is an alternative to grilling on coals – the fish is cut along the back, peeled, and its belly is filled with butter or garlic oil. The fish is spread on a baking sheet, poured with a small amount of water, and baked until the water boils away – thus, a very delicious crispy crust is formed on the carcass.

Fishery in the Ionian Sea

Fishery in the Ionian Sea – most of the Greeks of the Ionian Islands and also tourists spend their leisure time on the seaside or at the sea. One of the popular activities is fishing. Greeks, inhabitants of coastal villages and towns, have been fishing since time immemorial. You can fish from the shore, from a boat, and underwater. The best places for fishing from the shore are rocks, stones where sea inhabitants live, and the best time is night. There are several ways of sea fishing from a boat: (1) the most common method is vertical jig fishing (jigging) in the open sea. It is recommended to use an echo sounder and a GPS receiver.

(2) Trolling: fishing from a moving boat, which allows the angler to cover a wide area of the sea, while the bait takes on a natural look. The boat speed and lead weight can be adjusted so that the bait moves to the desired depth. Usually this method is used to catch marlin, wahoo, sailfish, and tuna.

(3) Spinning: artificial bait is thrown where you think the fish is. After that, the bait is allowed to descend to the desired depth and begin to wind the line back on the reel so that the movements of the bait copy the movements of the fish.

(4) Drifting: fishing from a drifting boat. In this case, the boat engine can be turned on, but so that it moves forward a little. The most important parameter when drifting is the speed of the boat in relation to the tide. If the wind and tide are in the same direction, the boat will drift too fast. Then the movement of the boat can be slowed down with a floating anchor.

(5) Fly fishing: during bottom fishing, the boat is at anchor in a fairly shallow place, and the bait is lowered to the seabed and held there with a suitable sinker. This method is used to catch large fish that feed on the bottom (common stingray, halibut). The very clean and warm Ionian Sea, in which there are no strong underwater currents, and the inhabitants of the marine world do not pose a danger, create excellent conditions for underwater fishing. Among the underwater rocks and in the cave of the Ionian Sea, there are octopuses and shrimps, sea urchins and sponges, morays and eels, sea basses, sea breams, giltheads, rofos fishes, stira fishes, mullets.

Flathead grey mullet (*Mugil cephalus*)

Flathead grey mullet (*Mugil cephalus*) – marine schooling mobile fish of the mullet family (mugilidae). The most famous subspecies of mullet. The body is fusiform, strong, and the sides are flattened. The head is large and broad. Length is up to 75 cm, and weight is up to 6 kg, but the weight of the catched fish is usually 0.3 kg. Reaches sexual maturity at the age of 6–8 years with a length of 30–40 cm. Spawning in batches in May–September, far from the coast. Fertility is 2–7 million eggs. Eggs of the fish are pelagic. During spawning it keeps in small flocks. It feeds on fouling, detritus, worms, and small molluscs. In winter it feeds on the open sea, in spring and summer – in the coastal zone. The meat is tasty, easily separated from the bones (Fig. 3).

Fig. 3 Flathead grey mullet. (Source: https://pbs.twimg.com/media/C4F0kMIWIAAkZEr.jpg)

Fountain of Neptune

Fountain of Neptune – The famous sculpture of Italy of the fifteenth century (1557), created by the sculptor Giovanni Angelo Montorsoli, installed in Messina. The original sculptural decoration of the spring has long been kept in the city museum. Currently, the fountain is decorated only with copies (Fig. 4).

The artistic idea is based on the legend that Neptune was identified with the Greek god Poseidon – the god of the seas and streams who separated Sicily from the Apennine Peninsula with a blow of a trident. The figure of the sea god is depicted at the moment when he calms the waters of the Strait of Messina with a single movement of his hand.

Fig. 4 Fountain of Neptune. (Source: https://img1.arrivo.ru/c4ca/63/12777/0/amipreside.jpg)

G

Galaxidi

Galaxidi (*Galaxeidi*) – a city-port in the north of the Gulf of Corinth on the southern coast of the large bay of the same name of the Krisseos Gulf, Greece. In recent history, Galaxidi was one of the main harbors of Greece. A fleet of more than 400 two- and three-masted schooners and caïques ("Turkish boats") was assigned to it, which carried goods to Great Britain. However, after 1890, prosperity ended. There are many Venetian style mansions built by shipowners in the nineteenth century. The Museum of Navigation and History is opened here.

Galleass

Galleass (*ital. Galeazza – large galley*) – a sailing and rowing vessel of the large galley type. For the first time in history, they were used by the Venetian fleet in the fourteenth century. In the sixteenth to seventeenth centuries they were part of the navies of many European countries. Venetian galleasses of the 1500s had a length of up to 50 m, a width of up to 8 m, a side height of 3.2 m. The crew of such a vessel consisted of a captain, two pilots, four officers, a boatswain with an assistant, a priest, 62 sailors and helmsmen, 268 rowers, and 254 soldiers and bombardiers – about 600 people in total. For the first time the Venetians used galleasses in 1571 in the naval battle of Lepanto against the Turkish fleet. At the end of the sixteenth century the galleasses were up to 80 m long, up to 16 m wide, 3 masts with 3 oblique sails, on each side – 1 row of oars (up to 10 rowers), in the stern – 2 large steering oars, 2 decks (lower – for rowers, upper – for soldiers and guns). It was armed with up to 70 guns of various calibers (Fig. 1).

Galleys were built with an elongated boarding "beak"; a topside was created on its nose in the form of a tower, where a catapult was installed. Sometimes the same tower was created aft. A wooden parapet protecting the rowers was placed along the side. The Galley was equipped with cannons: bombard, which appeared in the 1370s, with culverins, falconets, etc. Experienced sailors were recruited in Venice, Dalmatia, and Greece. There were no slave-rowers on Venetian Galleys, the rowers were free-born, both Venetians and Dalmatians. Galley team consisted of convicts, slaves, and volunteers. Gaff-sails were used as an auxiliary propulsor for Galley, which were set on 1–2 masts.

Gallipoli

Gallipoli – (1) a small island in the Gulf of Taranto of the Ionian Sea, which is part of the administrative region of Apulia, Italy. The island is high and rocky. It is connected to the mainland by a stone

I. S. Zonn, *The Ionian Sea Encyclopedia*, Encyclopedia of Seas,
https://doi.org/10.1007/978-3-031-08206-1

Fig. 1 Galleass. (Source: https://thelib.ru/books/00/11/42/00114253/i_009.jpg)

bridge, therefore it is sometime called peninsula. The bay of Sano-del-Canneto juts out into its eastern shore. According to myths, Gallipoli was founded by the Greek hero Idomeneus and for a long time was part of Ancient Greece, then became part of the Roman Empire, and later the Byzantine one. During the Middle Ages, the island was ruled by the Normans. 1268 to 1300 here ruled Charles I of Anjou, and then Gallipoli entered the princely territories of Taranto. In the sixteenth century the island was conquered by the Spaniards. From 1860 the island became part of the united Italy. On the island is the city of Gallipoli.

(2) From Greek "Beautiful city," located on a small island of the same name, on the coast of the Ionian Sea. In fact, the island is the historical part of the city, and the modern city is separated by a narrow sea strip and connected with the island by a bridge. For the construction of most of the old buildings, cathedrals, and churches, local rocks of snow-white limestone were used. The city presents masterpieces of Baroque architecture of the sixteenth to eighteenth centuries. These are the Church of St. Francis of Assisi, the Cathedral and Convent of St. Domenico, the Cathedral of St. Mary, Palazzo Pirelli, and the Cathedral of St. Agatha, which contains a real gallery of Baroque paintings by Salentine and Neapolitan artists.

(3) A port, located in the eastern part of the Gulf of Taranto of the Ionian Sea, 6.3 km north-north-west of Cape Pizzo on the northern coast of the high and rocky island of Gallipoli, which is connected to the coast by a stone bridge. It was built by Ferdinand I of Bourbon. The bridge later became an important commercial hub. A valuable and necessary cargo was sent from here – olive oil for lamps, which was produced on churns in Gallipoli. Until now, some of these churns have survived in Gallipoli, which were equipped in basements. It was convenient to maintain the required humidity and temperature in them, which increased the acidity of the olives, since oil of high acidity is needed for lamps. The resulting oil was of exceptional quality, which was sold at a high price to England, France, Spain, and Germany. Gallipoli was not subject to taxes and duties, therefore, the city's well-being remained at a high level for almost three centuries, until a new product appeared – oil (Fig. 2).

Oil production began to decline, basement mills were closed, churns were built on the surface, however, for the production of edible olive oil. By the way, Apulia has succeeded in this too – olive oil from Apulia with a bright aroma is considered one of the best. The port has three harbors: for vessels with a length of 130 m, for fishing vessels and for boats.

Garfish (*Belone belone*)

Garfish (*Belone belone*) – schooling, pelagic fish of the *Belone* genus, Belonidae family. The body is elongated, ribbon-like, very flattened from the sides. The back is dark, bluish-green. The belly and sides are silvery, and the jaw is dark, rather black. Spawning is portional and occurs at depths of up to 20 m. Fry keep in the coastal zone, and by the end of the first year they roll back to depth. Adult fish feed on small fish and invertebrates. The largest catches are the Italian coast (Fig. 3).

General Fisheries Commission for the Mediterranean (*GFCM*)

General Fisheries Commission for the Mediterranean (*GFCM*) – the agreement on the commission was approved by the FAO Conference in 1949, in accordance with the provisions of Article XIV of

Fig. 2 Gallipoli. (Source: https://www.comune.gallipoli.le.it/images/notizie/anno-2019/gallipoli-1.jpg)

Fig. 3 Garfish. (Source: https://a.d-cd.net/xGVA1PjLphdUk8srhcsL53jhFQw-960.jpg)

the FAO Constitution, and commenced in 1952 (amended in 1963, 1976, and 1997). The latest amendments were related to the change of the previous name – "General Fisheries Council for the Mediterranean." The Commission includes 24 members: 22 countries of the Mediterranean and Black Sea region, 1 non-Mediterranean country and the EU – Albania, Algeria, Bulgaria, Croatia, Cyprus, Egypt, France, Greece, Israel, Spain, Lebanon, Libya, Malta, Montenegro, Monaco, Romania, Morocco, Slovenia, Italy, Syria, Tunisia, Turkey, and Japan. The purpose of the Commission is the development, protection, rational management of living marine organisms, their best use and sustainable development of aquaculture. The main functions of the Commission: coordination of government efforts to effectively manage fisheries at the regional level, in accordance with the FAO Code, the acceptance of binding recommendations for the conservation of fisheries and management in its conventional area, strengthening fisheries management in the region, cooperation with other regional fisheries organizations. The Commission holds annual regular sessions. It implements policies and activities through the Secretariat, located at FAO headquarters in Rome, Italy. During the intersessional period, it works through the Scientific Advisory Committee, the Aquaculture Committee, and the Conciliation Committee.

Geographic Sub Areas of the Ionian Sea

Geographic Sub Areas of the Ionian Sea – The Mediterranean Sea is divided into 30 subareas known as geographic sub areas (GSA). Such areas represent a compromise in terms of legal, geographic, and economic aspects. The term "sub" refers to the fact that the Mediterranean Sea is one of 60 large marine ecosystems on Earth. Geographic subareas in the General Fisheries Commission for the Mediterranean (GFCM) are established by the GFCM Resolution 31.02.2007 in agreement with the Scientific Advisory Committee (SAC). The 30 subregions differ greatly in size and characteristics. The geographical division of the fishing areas in the Mediterranean is still in development and is subject to periodic review by the SAC. Of the 30 subregions of the Ionian Sea, it is divided into the Western Ionian sea – number 19, Eastern Ionian Sea – number 20, and Southern Ionian sea – number 21.

Gerogombos

Gerogombos (*Gerocombos*) – the western tip of the Kefalonia island. It is a cape where two seas, the Ionian and the Adriatic seas, coalesce. The cape is composed of high cliffs. There is an assumption that there was a palace of Odysseus. In the village of Chavriata – this is the "balcony of the Ionian Sea" – the westernmost lighthouse of the island of Kefalonia is installed in the form of a tetrahedral stone tower. It helps sailors to pass sharp and dangerous rocks of Kefalonian shores. Lighthouses are an important part of Greek maritime history. They are symbols of hope and security.

Geyden Loggin Petrovich (1772–1850)

Geyden Loggin Petrovich (Lodewijk Sigismund Vincent Gustaaf Reichsgraf van Heiden) (Geiden Login Petrovich) (1772–1850) – admiral. Born in Holland. He began his service in the Dutch Navy. By 1794 he rose to the rank of lieutenant. In November 1795 he entered the service of Russia and was appointed lieutenant commander in the Black Sea Fleet under the name Loggin Petrovich. In 1796 he was at the Nikolaev port, in 1797 he commanded the boat "Christopher," in 1798 the commander of the brigantine "Alexei" sailed in the Black Sea in the squadron of Vice Admiral P.I. Lezhnev. Participated in the campaign of the squadron of F.F. Ushakov to the Mediterranean Sea in 1798–1800. In 1799, commanding the brigantine "Alexei," sailed in the Mediterranean Sea and participated in the landing of troops in the city of Otranto, after which he returned with the fleet to Sevastopol. He commanded the frigate "John Chrysostom." (Fig. 4).

In 1801, with the first group of midshipmen sent from the Marine Corps, he sailed on the frigate "John Chrysostom" through the ports of the Black Sea, showing future officers the naval theater. In the next campaign on the same frigate he sailed between Constantinople and Nikolaev and surveyed the shores. In 1804 he was promoted to captain second rank with transfer to the Baltic Fleet. Appointed as a member of the prize committee and assigned to the Marine Corps with the rank of major. He commanded the ship "Conception of St. Anna" in the Baltic Sea. From 1805 to 1808 he was the head of the economic expedition for the crew department. After the Russian-Swedish war broke out in 1808–1809, he was promoted to captains of the first rank and assigned to the rowing fleet, which was trained in St. Petersburg and sent in detachments to Finland. At the beginning of July, having united the forces of three detachments (40 ships), he decided to bypass the Jungfersund along a narrow strait between Kimito Island and the mainland. On July 9, he led the battle at Kimito against 25 ships of the Swedish rowing flotilla and forced

Fig. 4 Russian Admiral Geyden L.P. (Source: http://www.navy.su/persons/04/images/geyden_lp_00.jpg)

the enemy to retreat to Sando. He was wounded in battle. He recovered by the fall and, commanding three squads of the rowing Finnish flotilla operating near the Åboland skerries, took part in battles with the Swedes. From 1809 to 1813 he commanded a rowing flotilla stationed in Finland. Promoted to captain-commander. In 1814 and 1815 commanded a flotilla in Åbo. In 1816 he was appointed the chief commander of the Sveaborg port and military governor, for the distinction he was promoted to rear admiral. From 1823 to 1826 he served in St. Petersburg at the shore. In 1826, in the squadron of Admiral R.V. Crown sailed from Kronstadt to the coast of Scotland. In 1827 he commanded the rearguard of the fleet during maneuvers at Krasnaya Gorka. In the summer of 1827, in the squadron of Admiral D.N. Senyavin, he went to England and was sent with a separate squadron to the Mediterranean Sea to prevent further massacres of the Greeks by the Turks. After joining with the British and French squadrons, on October 8, 1827, he participated in the defeat of the Turkish-Egyptian fleet at Navarino. In 1828 he was appointed to command the second Division of the Baltic Fleet. In fact, he headed a squadron located in the Mediterranean Sea. During the Russo-Turkish war of 1828–1829 blocked the Dardanelles. In 1830, commanding a squadron of seven ships, he was sailing by the coast of the seas. After returning to Kronstadt, he was appointed to command the first Fleet Division. In 1833 he was promoted to admiral. In 1834 he was appointed military governor of Revel, and in 1838 concurrently also a chief commander of the Revel port. An atoll in the Marshall Islands archipelago, a bank in the Bristol Bay of the Bering Sea, USA, and a mountain in Kamchatka are named after Geyden.

Giardini

Giardini – port at Capo Schiso cape, south of Taormina, Sicily, Italy. It is protected from eastern winds by a breakwater 150 m long. On the western side of the breakwater, embankments are built: depths along them are 4.6–8.6 m.

Giardini-Naxos

Giardini-Naxos – one of the richest resorts on the island of Sicily. It is situated in close proximity to the city of Taormina on the island of Sicily, Italy. Back in the eighth century BC the ancient Greeks – immigrants from Chalkis – landed in Sicily exactly in the place where the city of Giardini-Naxos stands today. From that moment, its story began. At first it was an ordinary Greek colony, based on the former islanders of Euboea. Quite quickly, the island of Naxos, located nearby, joined the developed area. In memory of this event, the town bears the double name given to it in 1978. Although, despite the seeming calmness of the area and the small size of the city, the wars left their mark here too. During the war of Sparta against Athens at the beginning of the fifth century BC Giardini-Naxos was almost

Fig. 5 Giardini-Naxos, a resort on Sicily. (Source: https://www.cruisegid.ru/assets/gallery/961/14806.jpg)

destroyed. In the sixteenth century the fort of Schisò Castle was built on the coast, which soon turned into a real defensive fortress that protected all the coastal villages of this area. A castle with Vignazza Tower remained from it. It was rebuilt again. Nowadays Giardini-Naxos lives off the tourism business. However, two centuries ago, iron forging and the creation of ceramic products were exclusive occupations here. These traditional crafts also remain the hallmark of the settlement. Nowadays, the museum contains many items created by the craftsmen of Giardini-Naxos. The historical museum itself with the ruins of the temple of Aphrodite is situated in an old fortress, erected about four centuries ago. From the city you can take a funicular to the city of Taormina (Fig. 5).

Gilt-head bream

Gilt-head bream – q.v. Dorado (*Sparus aurata*).

Glafkos

Glafkos – river in Greece. It flows into the Gulf of Patras of the Ionian Sea. The first hydroelectric power plant in the country was built on it. It is currently open as a museum.

Gramvousa

Gramvousa – the common name of two uninhabited islands: Agria Gramvousa and Imeri Gramvousa, Greece, near the northwestern tip of Crete, which was once a haven of pirates. It is considered to be the confluence of the three inland seas of the Mediterranean Sea – Ionian, Cretan, and Libyan. Sea water here, specifically in the Balos lagoon, depending on the time of day, has from 12 to 17 shades: from light emerald to deep blue, while remaining always crystal clear and transparent.

Great Greece

Great Greece (*лат. Magna Graecia*) – a historical area with a complex of ancient Greek settlements – colonies founded in the antique period, which included part of the territory of modern southern Italy (the Apennine Peninsula and Sicily). By the third to second centuries BC, it was annexed by the Roman Republic (Fig. 6).

It emerged during the Second or Great Greek colonization in the eighth to sixth centuries BC. The settlements were withdrawn from the mainland and insular parts of Greece and Asia Minor on the Mediterranean coast. The reasons for colonization were overpopulation, low agricultural productivity, and the growth of the nobility land holdings at the expense of land-hungry peasants. The new Greek colonies were founded primarily as agricultural settlements; however, trade became the predominant occupation of their population. The precondition for the emergence of numerous cities, often far from each other, was the success of the art of navigation. Great Greece is one of the three directions (western) along which the Great Greek Rhegion, Lokroi (Epizephiroi), Kaulonia, Scylletium, Kroton, Petelia, Sybaris, Herakleia (Lucania), Metapontion, Taras, and Kallipolis.

Greater Pipefish, *Syngnathus acus*

Greater Pipefish, *Syngnathus acus* – fish, belongs to the Syngnathidae family. It has an elongated segmented body, reaching a length of approximately 35 cm.

Greater weever

Greater weever (Trachinus draco) – a venomous marine fish of *Actinopterygii* (Trachinidae) family, one of the most venomous marine dwellers in the temperate zone. The fish of the Ionian and Mediterranean seas is black in color, and the body length does not exceed 45 cm. It dwells on the muddy or sandy bottom of bays and gulfs. It feeds on small fish, worms, and crustaceans. If there is danger, this fish first makes a warning in the form of an expanded fin, and, if necessary, jumps out of its shelter and pierces the enemy with a poisoned thorn. Due to its subterranean lifestyles, aggressiveness, and high toxicity, the greater weaver is especially dangerous for swimmers near the shore, walking barefoot in shallow water, as well as for fishermen. It has a very strong venom similar to that of a snake. People affected by the greater weaver

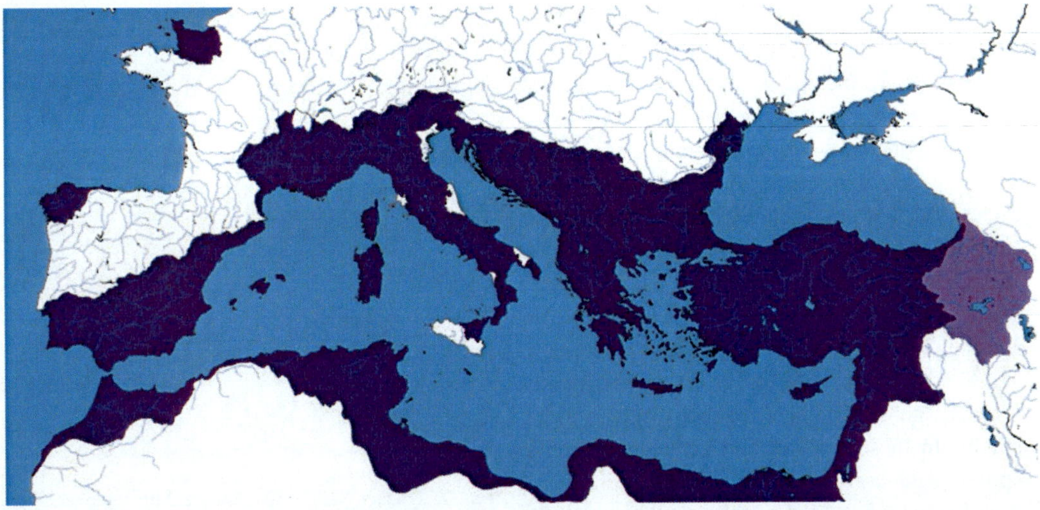

Fig. 6 Great Greece. (Source https://i.redd.it/3z983li0r0621.png

have inflammation and swelling of the affected areas, loss of consciousness, vomiting, convulsions, and an irregular heart rhythm. Some cases are fatal. In order to destroy the poison at first aid, it is necessary to inject a solution of potassium permanganate into the wound. Although they inhabit the Ionian Sea, they do not like noise and do not appear near the resorts.

Greater Weever (*Trachinus draco*)

Greater Weever (*Trachinus draco*) – a species of carnivorous ray-finned fish of the weever fish family (Trachinidae). There are five types of Greater weevers, which are found off the coast of Europe, the northern and western coasts of Africa, as well as off the coast of Chile. It is elongated with angular big head, bottom fish.

The length of the fish's body is 40 cm and it weighs about 300 g. It reaches sexual maturity at the age of three. They live mainly in bays and coves with a flat sandy or oozy bottom. Usually it is buried in soft ground, trapping prey in ambush – small fish and crustaceans, high-set eyes remain above the bottom surface. The lower jaw is longer than the upper jaw, and the mouth with small conical teeth is directed obliquely upwards. Eyes are located on the upper side of the head. These fish have two dorsal fins: the first one is short, it has 5–7 spines, and the second dorsal and anal fits are long, and they have 21–24 rays each. The ventral fins are located on the throat. The scales are very small, cycloid, located in oblique rows. They feed on small fish, worms, and crustaceans. The first dorsal fin of a greater weever has an intense black color that differs sharply from the pale yellow and brown tones of the rest of the body parts and the surrounding sand. Greater weever is the most dangerous poisonous fish in Europe. The long sharp spikes of the operculum and the first dorsal fin are covered with skin, only their tips stand out. There are deep grooves on the spikes. In these furrows and at the base of the spikes there are poisonous glands. There is no duct for poison. Probably, when a thorn prickles the flesh, the glandular cells are broken, and the poison is released and is introduced along the grooves into the wound, like a syringe needle. The poison of greater weevers is very strong. When you are injected with a poisonous spike, a painful, sharp, burning, or stabbing pain bursts out, which lasts for several hours or even a whole day, if there will be no treatment. The affected limb is inflamed and severely turgescing. Other symptoms include loss of consciousness, heart palpitations, slow heart rate, fever, headaches, delirium, severe vomiting, seizures, and rough breathing. In severe cases, the victim may die. Morphine does not usually give relief.

Greater Yellowtail, *Seriola dumerili*

Greater Yellowtail, *Seriola dumerili* – body slender, tall, and thin with a torpedo-shaped form; the profile of the head forms a wavy line, the caudal fin with a deep excision. The anal fin is shorter than the second dorsal fin, and in both the first rays are longer than the others. In the first dorsal fin 5–7 prickly rays, which are connected by a membrane. Pectoral fin is short. The back and upper part of the body are dark blue with a purple tint, the sides of the body are yellow, and the belly is pale yellow. It reaches a length of 190 cm and a mass of 50–60 kg. It dwells near the surface of the water, at a depth of 20–70 m, in small flocks, often in fast currents near rocks and islands. Greater yellowtail is a powerful swimmer, making significant migrations every year (Fig. 7). The largest

Fig. 7 Greater Yellowtail. (Source: https://biogeodb.stri.si.edu/caribbean/resources/img/images/species/3654_6171.jpg)

species usually lead solitary lives, and the youngsters often keep in small flocks under the dome or near the jellyfish. Greater yellowtail is widespread in the tropical and warm waters of the Atlantic Ocean; in the waters of Northern Europe, this species is extremely rare. Meat has an average value.

"Greece pivot of the western coast"

"Greece pivot of the western coast" – volume 1, Western coast, 1971.

Greece, Greece Republic

Greece, Greece Republic – is a state in Southern Europe. It is situated on the Balkan Peninsula and numerous islands. The population is about 10.7 million people. It is washed by the Aegean (including the Ikarian Sea) and Thracian seas in the east, the Ionian Sea in the west, and in the south by the Mediterranean and Cretan seas. It has a land border with Albania in the north-west, Macedonia and Bulgaria in the north, with Turkey in the north-east. The area of the territory is 132 000. km^2, including the area of islands – 25.100 km^2. Greece includes about 2,000 islands, which account for 20% of the entire country. Greece's heavily indented coastline stretches for 4,000 km, and with islands for more than 15,000 km. The territory of Greece is divided into three parts. (1) Mainland Greece, which includes Greek Macedonia (northern part of Greece), Thrace (northeastern part), Epirus (northwestern part), Thessaly (plain region in the east, which is washed by the Aegean Sea), and Central Greece (region around Athens). The Peloponnese is the largest peninsula in Greece, connected to the mainland by a narrow Isthmus of Corinth, through which the famous Corinth Canal was dug. (2) Insular Greece includes Crete – the largest island in the Mediterranean Sea, the Ionian Islands (Zakynthos, Kerkyra (Corfu), Kefalonia, Levkas, Ithaca, etc.), located to the west of mainland Greece; Saronic Islands (Salamina, Aegina, Poros, Spetses, etc.) in the gulf separating Attica and Peloponnese. (3) Islands of the Aegean Sea: Euboea (3654 km^2), which is connected to the mainland off the east coast by a bridge over the Euripus Strait; Lesbos (1630 km^2), located off the coast of Turkey. Northern Sporades (Skyros, Skopelos, Gioura, etc.) – in the northwestern part of the Aegean Sea. The Cyclades is an archipelago in the center and south of the Aegean Sea (islands: Amorgos, Andros, Naxos, Kythnos, Mykonos, Santorini). Southern Sporades (Dodecanese) is a group of 12 islands in the southeastern Aegean Sea off the coast of Turkey. The largest islands are Rhodes, Samos, etc.

The Greek landscape is an alternation of rocky, usually treeless mountains, densely populated valleys, numerous islands, straits, and bays. Picturesque cliffs, beaches, and exotic grottoes provide great opportunities for sea recreation and mountain tourism. The widespread occurrence of limestone, especially in the western part of the country, led to the formation of sinkholes and caves. Mountain ranges occupy almost a quarter of the country's surface. Medium-altitude mountains predominate (up to 1200–1800 m). The highest point in Greece is Mount Olympus (2917 m). Pindus, Parnassus, the mountain range of Central Greece, and Taygetus also rise above 2000 m. There are few plains. They are concentrated in the northeastern and eastern half of the country, except the Peloponnese, where plains prevail on the west coast. There are strong earthquakes in Greece. This is an area of high seismic activity. In terms of the frequency and strength of earthquakes, Greece is in second place in Europe after Italy. There are also volcanic phenomena (volcanoes Santorini and Nisyros) (Fig. 8).

The climate can be divided into three types: Mediterranean, Alpine, and Moderate, each of which affects a specific area. The Pindus mountain range (up to 1500 mm of precipitation) strongly influences the climate of the mainland of the country: the regions located west of the slopes of the Pindus (Epirus) receive more

Fig. 8 Greece. (Source: https://d530gosedmutm.cloudfront.net/landing-pages/c8cc86c7206548f1a5278afb28e1a805.jpeg)

precipitation (900–1200 mm) than the regions located on the eastern side of the ridge (Thessaly) (400–700 mm). Almost all precipitation occurs in the autumn-winter period. Spring and summer are dry. The average January temperature in the north is + 4° to + 6°C, in the south –12° to –14°C, in July, +25 to +27°C, respectively. Rivers are few, mostly short, mountainous, with sharp fluctuations in flow rates and winter floodtimes. In the east, many of them dry up. The most significant rivers: Achelous, Nestos, Strymónas (or Struma), Vardar, and Alkmon. They are small in length and fast flowing, since most of them are mountainous. In mainland Greece, the rivers Arachty, Acheloos, Haliacmon, Pineios, and Spercheios, known from Greek myths and legends, originate in the Pindus mountains and flow either into the Ionian Sea or the Aegean Sea. Most of the Peloponnese is irrigated by the Alfeios and Eurotas rivers. In the summer, the runoff is sharply reduced, the rivers become shallow, and in some places dry up. Rivers are not suitable for navigation; however, they play a great role as energy sources. Floods are the most common natural disaster in Greece. They occur regularly and are caused by heavy rains. There are more than 20 lakes with area varying from 10 to 160 km^2 (the largest is Prespa) and mineral springs. Due to the broad distribution of limestones, shallow soils prevail. It is believed that in the past Greece was densely forested; however, most of these forests were cleared at the beginning of antiquity. As a result of the destruction of forests, the mountain slopes have undergone severe erosion. The outstanding Greek mastermind Plato wrote in "Critias" that Greece is the backbone of a body emaciated by ailment, when all the thick and soft earth was washed away, and thus only a devoured skeleton remained from the country. Mountain-forest brown soils prevail, mountain-forest soddy-calcareous soils are widespread in the central part. Brown soils of dry forests and shrubs are common. The natural vegetation of Greece has been significantly disturbed by human activity, and forests have been particularly severely degraded. Currently, they occupy only 20% of the country's area, mainly in the north and in the Pindus mountains. In the lower parts of the mountain slopes, they consist of evergreen stiff-leaved shrubs, higher – of deciduous species (shiblyak), low, thorny shrubs (maquis) are widespread. At altitudes from 1000–1700 m, broadleaf forests of beech, oak, and chestnut are often found, above 2000 m – conifers from Bosnian pine and Greek fir. Cypresses and plane trees are common. Some are several thousand years old. The olive is very widespread – one of the most valuable trees in Greece and the entire Mediterranean. The Greek mountains are very rich in various types of herbs and flowers. The plateaus of Crete and the Peloponnese are covered with a carpet of thyme, sage, hyacinths, irises, etc. Not many species of wild animals have survived in Greece. Their populations are small, although they are very diverse. This is due to the fact that for over 8000 years people have been actively exterminating them. Small animals are typical here: hares, badgers, porcupines, and various types of mice. Reptiles are numerous – snakes and lizards. Of the large mammals, the most common are the brown bear (rarely), jackal, fox, lynx, badgers, hares, and wild boar. Many species of animals are listed in the Red Book, including the sea turtle caretta and the monk seal. Among the birds, very common are wild ducks, kingfishers, and partridges, as well as predators – owls, eagles, and kites. There are many seagulls in the coastal areas. In the waters of Greece, there is a huge variety of shellfish, many octopuses, jellyfish, and fish, although the stock of the latter has significantly decreased in recent years. The country has significant reserves of minerals (magnet, bauxite, nickel ore, magnesite, marble, etc.). The world's largest emery deposit is located on the island of Naxos. Oil and natural gas deposits have been found in the Aegean Sea. Greece is considered the cradle of Western civilization. In the period of about 3000 years BC, on the island of Crete, a well-developed Minoan civilization arose, the culture of which subsequently spread to the mainland. It was followed by the era of the Cretan-Mycenaean or Aegean civilization. Later, Greek city-states arose, as well as the ancient colonies of Magna Graecia. The cultural level of development extended to the entire Mediterranean region, which was reflected in architecture, theater, science, and philosophy. City-state Athens and Sparta played a leading role in the victory over Persia; however, later they themselves were defeated by Thebes, and later by the Macedonian kingdom. The latter, under the leadership of Philip II of Macedon and his son Alexander the Great, reached extraordinary power, which was a harbinger of the beginning of the Hellenistic era. However, Macedonia was destroyed by the Romans in 146 BC, and Hellas became dependent on the Roman Republic. Further mutual influence of the Hellenic and Roman cultures is formalized in the culture of the Byzantine Empire. It remained the main cultural center for a thousand years, until its fall under the onslaught of the Ottoman Empire on May 29, 1453. During the Ottoman rule, there was a system of Ottoman millets that helped Orthodox Greeks to preserve their traditions for four centuries and fostered their cohesion based on religion, which played an important role in the forming of modern Greek identity. Greece gained its independence from the Ottoman Empire in 1830 after the national liberation war of 1821–1829. The first president of independent Greece was Ioannis Kapodistrias). However, soon a monarchy was established in Greece and a minor Otto of Bavaria from the Wittelsbach dynasty was invited to the throne. An uprising in 1843 forced King Otto to present a constitution and establish a representative National Assembly. However, in 1863, Otto was overthrown, and the Danish Prince William was invited to the Greek throne, who became George I. In honor of his coronation, Great Britain presented Greece the Ionian Islands. In 1877, at the initiative of Charilaos Trikoupis, the most prominent figure in Greek politics at the time, the king was deprived of the right to influence the National Assembly by the advancement of vote of non-confidence to the country's prime minister (Fig. 9).

After the end of the Balkan Wars of 1812–1913. Greece has significantly expanded its territory. In the years that followed, the political struggle between King Constantine I and Prime Minister Eleftherios Venizelos split Greek society on the eve of World War I. After the end of the latter, Greece entered the war with Turkey, then led by Mustafa Kemal Ataturk. This war led to the loss of a part of the territory of Greece and a massive exchange of population between the two countries within the framework of the Treaty of Lausanne signed on July 24, 1923. On October 28, 1940, fascist Italy demanded that Greece provide a springboard for the deployment of its forces, to which Prime Minister Ioannis Metaxas replied with a categorical "no," realizing that now the war for Greece was becoming inevitable. Although German troops were withdrawn from the territory of the country in 1944 as a result of the offensive of the Soviet army in the direction of Yugoslavia, since October 28, 1942, Greece has been celebrating the day of the answer "No" to the ultimatum of the Axis countries of Italy and Germany as a national holiday – the Ohi Day or Oxi Day (from greek – "no"). In the Civil War of 1946, the Greek Communist Party was defeated. In

Fig. 9 Athens. (Source: https://www.grekomania.ru/images/greek-articles/tourism/big/331_athens24-1.jpg)

1949, the monarchy was restored in Greece, which was finally abolished on April 21, 1967, after the military coup d'etat of the "black colonels," supported by the United States. After the overthrow of the military junta in 1975, a new constitution was adopted. The previous Prime Minister Konstantinos Karamanlis returned to the country from Paris, and as a result of a popular referendum, the monarchy was abolished and Greece became a parliamentary republic. In opposition to Karamanlis New Democracy, Andreas Papandreou founded PASOK. These parties still remain the most influential in the country. In 1980 Greece re-joined the NATO military alliance (in 1974 it withdrew from it in protest against the occupation of northern Cyprus). Greece became a member of the European Union on January 1, 1981. On the Macedonian issue, Greece still remains adamant in its decision not to recognize the Republic of Macedonia for the latter's attempt to declare its rights to the historical and cultural heritage of Macedonia. In December 2008, riots broke out in Athens, which quickly engulfed and rocked all of Europe. The cause of the unrest was a long-standing dissatisfaction with the economic situation, which was greatly aggravated by the World Economic Crisis that began in the same year. The reason for the deployment of mass protests, which often escalated into pogroms and riots, was the murder on December 6 of a 15-year-old teenager by an Athenian patrol officer. Since the spring of 2010, almost ceaseless nationwide strikes, riots, and terrorist attacks have become a real tentamen for the government. Early parliamentary elections were held on January 25, 2015. The victory was won by the Coalition of the Radical Left (Syriza), for which 36.34% of voters voted. Modern Greece is the heir to the culture of Ancient Greece, considered the cradle of Western civilization, the birthplace of democracy and Western philosophy, basic physical and mathematical sciences, theater, and the modern Olympic Games. Its rich cultural heritage and geographic location make Greece one of the most visited countries in the world. Greece is a unitary state consisting of 13 administrative districts – periphery, each of which is divided into prefectures (nomoi). In 1983, it was legally established that councils elected by the population through direct elections are in charge of local issues. The current Constitution entered into force on June 11, 1975. According to the form of government, Greece is a parliamentary republic. The political regime is democratic. Greece is an industrial and agrarian country with a chaotically developing economy. For a long time, it has been considered one of the most developed agrarian states in Europe. GDP from agriculture is 8.3%, from the industrial area – up to 27.3% of services – over 64.4%. At the same time, the needs of citizens in liquid fuel are covered only by imports. The sectoral structure is developed unevenly and disproportionately throughout the territory. The following areas of economy are developed: mining industry – mining of nickel ores, bauxite, emery, magnesite, pyrite; metallurgical steel production, cement, electrical engineering and power engineering, petrochemical, some types of mechanical engineering, production of building materials; the light (textile) industry is considered as a priority – the country's economy replenishes by almost 19% from the industry, while more than 21% of the population is involved in it, and the food industry. Greece is one of the leading maritime powers. Commercial shipping plays an important role for the economy. Greece has the largest fleet in the world – passenger, merchant, oil tanker. Greek shipping is intended for the state's foreign economic relations. Only in some countries, such as Russia, the USA, China, Turkey, and France, sea transport carries out interregional inland transportation. This transport carries more than 80% of foreign trade cargo. Greek shipowners own more than 3500 vessels of various types. Greek ships are usually registered under "flags of convenience," the crews are international. The merchant fleet ranks third in the world after the Liberian and Panamanian ones. As of 2010, Greek shipowners own 25% of the world's fleet. The Greeks control 22.3% of the world's tanker fleet, 18.1% of cargo ships, 9.4% of chemical transport, 4.5% of container ships, and 6.7% of ships. The qualitative improvement and modernization of the Greek merchant fleet goes on, and the average age of ships of which does not exceed 11.9 years. The merchant fleet provides approximately 20% of GDP. The fishing fleet in Greece is small and consists mainly of small vessels. There are few poly-size trawlers. The country's agriculture is based mainly on small private farms. About 30% of Greece's land is arable. It employs 12% (530 thsd.) of workers employed in the country's economy. The share of agriculture in Greece's GDP is 7% ($16 billion) of GDP. This area employs a large number of immigrants, mainly Romanians and Albanians. The country's agricultural sector is largely funded by the EU. Due to the geographical and climatic features in Greece, a huge variety of crops can be grown: from tropical (pineapples) and all types of citrus fruits to crops specific for the northern regions (berries, mushrooms, etc.). Only in the valleys of Thessaly, Thrace, and Macedonia wheat, corn, barley, and sugar beets are grown. Greece ranks first in Europe for the production of cotton and tobacco. Animal husbandry plays an insignificant role in the Greek economy, which is due to the small amount of forage and insufficient land plots. The mountainous terrain allows raising no more than 4.5 million goats and about 8 million sheep. The number of pigs and cows in Greece is even smaller (in total – about 1.8–1.9 million). Rabbits and poultry are bred in several regions of the country, exclusively for their own needs, selling a small part of it to third parties. Also mules, donkeys, and horses used for agricultural work as a pulling force are bred for themselves. The country's agricultural enterprises grow a number of crops for export. These include olives, tobacco, citrus fruits, grapes, and cotton. Greek olive oil tastes great and has won its market share. Greece's relatively dry climate necessitates irrigation. More than half of all arable land – 2.5 million hectares – has to be irrigated. At the same time, the yield of agricultural crops is low. The most important agricultural crops include tomatoes (they are grown per year over 1.7 million tons), potatoes, sugar beets, olives, legumes, and cereals. In good years, Greek peasants harvest more than 5 million tons of grain. Some barley, corn, and wheat are imported. Forestry is of secondary importance. Tourism in Greece is one of the main sources of foreign exchange earnings. Every year 20 million tourists come to Greece, which forms 15% of the country's GDP. Foreigners are attracted by the rich history, culture, and favorable climatic conditions. Over the past few decades, there has been a significant increase in beach tourism in Crete, the Peloponnese, Rhodes, and the Ionian Islands. The beginning of the twenty-first century was marked for the country by serious financial changes. Foreign investors flocked to Greece. On the one hand, this stabilized the economy and closed gaps in some important issues, on the other hand, the national system had to adapt to Western integration. As a result, Greece began to systematically yield to its partners in the European Union. Only multibillion-dollar loans from American, Italian, French, and German banks help maintain capital.

Greek Navy

Greek Navy (*Ellinico Polemico Naftico*) – one of the branches of the armed forces of the Hellenic Republic. The history of the modern Greek Navy is counted from the moment Greece proclaimed independence from the Ottoman Empire in 1828. From the beginning of the twentieth century, the creation of the Greek Navy is carried out on the principles of copying the British Navy – organizational structure, principles of training, communications, etc. borrowed from a European neighbor. Many ships for the Greek Navy were built in the UK. Among the main factors determining the maritime interests of Greece are: the persisting threat of a conflict with Turkey (over disputed island possessions, borders, territorial waters, control over the sea and air space in the Aegean Sea and the Cyprus problem); the need to protect and patrol the maritime economic zone and more than 3000 islands, control and protect maritime communications; dependence on the supply of energy raw materials by sea due to the poor development of the pipeline system for the transportation of petroleum products. The naval forces are considered a tool to strengthen its position in world politics by nation's leadership. The battleships of the Greek fleet are

Fig. 10 Greek Navy. (Source: https://www.rusdialog.ru/images/news/12df62faebed02271a6c92a8e4c512c0.jpg)

part of the permanent formation of the NATO Allied Navy in the Mediterranean Sea, regularly participate in the maneuvers of the operational formation "Euromarfor" created within the WEU, the main purpose of which is to conduct peacekeeping, humanitarian, search and rescue, and combat operations to resolve armed conflicts and crises (Fig. 10).

Greece also intends to join the Black Sea naval group "Blackseafor," an agreement on the establishment of which was signed in March 2001 in Istanbul by representatives of Russia, Turkey, Romania, Bulgaria, Ukraine, and Georgia. The functional purpose of the Greek fleet is to protect the national territory from aggression, ensure national security, as well as allied interests within the North Atlantic bloc and the European Union (EU), and the main tasks in peacetime are: maintaining a favorable operational regime in the zone of responsibility, protecting shipping, and conducting all types of reconnaissance. In wartime, the fleet must be ready to conquer dominance in the eastern Mediterranean, support ground forces in coastal areas, block the Black Sea straits, conduct amphibious operations, and protect communications. Currently, the Greek Navy includes three commands (fleet, logistics, and training), naval aviation, and three naval districts. In addition, in operational terms, the chief of staff of the Navy is subordinate to a regiment of marines and a squadron of patrol aircraft from the ground forces and the air force. The total number of personnel of the Greek Navy is 19,500 people, including 3,000 officers and 200 women. Officers are trained by the Naval Officers Academy, and petty officers – by a special school (Petty Officers School). The command also includes three training centers, a training ship "Aris" and three small training sailing vessels. In 2014, the Greek Navy included 8 submarines, 14 frigates, 15 missile boats, 12 gunboats, 6 patrol boats, 7 minesweepers, 12 landing ships, and 36 auxiliary ships. The main headquarters of the Navy is located in Athens. The main base of the navy is Salamis Island. The second naval base is located in the Souda.

Greek Revolution

Greek Revolution 1821–1892 (Greek War for independence) – the liberation uprising of the

Greek people against Turkish rule, which began in the fifteenth century. The secret revolutionary society "Filiki Eteria"), created in 1814 in Odessa by Greek patriots, played a significant role in the preparation of the uprising. The uprising raised by A. Ypsilantis (one of the leaders of "Filiki Eteria") in Moldova in February (April) 1821 gave impetus to the anti-Turkish uprising of Greece, which began in March (April) 1821 and within 3 months covered the entire Morea (Peloponnese), part of mainland Greece, the islands of Spetses, Hydra, Psara, etc. The Greek uprising grew into a national liberation revolution, the main driving force of which was the peasantry. In the summer of 1822, the 30,000-strong Turkish army invaded Morea, however, facing the fierce resistance, was forced to retreat. In 1821–1822 the rebels liberated a significant part of Greece. In January 1822, a national assembly was convened at Piada, near Epidaurus, which proclaimed independence in Greece and adopted the first Greek constitution, the Epidaurus Organic Statute, which declared the country an independent republic. A. Mavrokordatos was elected its president. In 1827, in the city of Troezen, I.A. Kapodistrias. There was no unity of government among the rebels. Thus, after the first successes contradictions intensified in their camp, which led to two civil wars (November 1823–June 1824, November 1824–early 1825), which significantly weakened the Greek liberation movement. In 1824, the Turkish Sultan Mahmud II drew his vassal, the Egyptian Pasha, Muhammad Ali, to the fight against the rebels, promising him the territories of Syria and Crete. In 1825 the Egyptian army devastated most of the Morea. On April 22, 1826, after an 11-month siege, Egyptian and Turkish troops captured an important rebel fort – the city of Missolonghi, and in June 1827 the Turks captured the Athenian Acropolis, after which only small centers of resistance remained in Greece. This development of events forced the leading European powers to actively intervene in the conflict. In many countries, the public came out in support of the uprising Greeks, and volunteers began to go to Greece. Russia, striving to strengthen its positions in the Balkans and the Mediterranean, initially supported the rebels. Great Britain, trying to prevent the strengthening of Russia's influence in the Balkans and at the same time striving to strengthen its positions there, in 1823, after the first successes of the Greeks, recognized them as a "belligerent party" and in 1824–1825 gave them cash loans. On March 23 (April 4), 1826, a protocol was signed in St. Petersburg between Russia and Great Britain, according to which the parties took on mediation in the Greek-Turkish conflict on the basis of granting Greece internal independence. According to the London Convention of 1827, France joined Russia and Great Britain in resolving the Greek-Turkish conflict. The Turkish side rejected the proposals of the three countries, after which Russian, French, and British naval squadrons were sent to the Peloponnese, which in the Battle of Navarino in 1827 defeated the Turkish-Egyptian-Tunisian fleet. The fate of Greece was actually decided by the Russo-Turkish war of 1828–1829, which ended with the victory of Russia and the conclusion of the Treaty of Adrianople in 1829, according to which the Ottoman Empire recognized the autonomy of Greece, subject to the payment of tribute to the Sultan. In 1830 Greece officially became an independent state (Fig. 11).

Greek submarine "Papanikolis (Y-2)"

Greek submarine "Papanikolis (Y-2)" ("*Papanicolis*") – one of the Greek submarines that successfully operated during World War II. "Papanikolis" together with the submarine "Katsonis" of same type made up the first type of submarines ordered by the Greek Navy after World War I. "Papanikolis" got its name in honor of the hero of the Greek Liberation War of 1821–1829, Dimitrios Papanikolis captain of the fireship. "Papanikolis" was built at the Chantiers de la Loire in Nantes, France, 1915–1927 and commissioned into the Hellenic Navy on December 21, 1927. The submarine took part in the Greek-Italian war of 1940–1941, carrying out six

Fig. 11 Greek Revolution 1821–1892, (Greek War for independence). (Source: https://etot-prazdnik.ru/wp-content/uploads/2019/03/maxresdefault-1.jpg)

war patrols in the Adriatic Sea. During one of them, on December 22, 1940, the submarine sank the small Italian diesel ship "Antonietta," and the next day, the 3952-ton troop carrier "Firenze" near Sazan Island. After the invasion of Germany, which came to the aid of the Italians in April 1941, "Papanikolis," together with other ships of the navy, left for the Middle East, from where, over the next years, it carried out nine war patrols. "Papanikolis" took part in two military campaigns in the Aegean Sea in 1942. During the first of them, between 11 and 14 June 1942, Papanikolis sank six small sailing ships, disembarked SOE agent in Crete, and took on board 15 New Zealand saboteurs. During the next campaign, between August 31 and September 15, she landed two Anglo-Greek sabotage groups on the island of Rhodes, which, during Operation Anglo, managed to attack two airfields on the island and destroy a large number of German and Italian aircraft. She carried out a war patrol in November, dropping people and supplies on the island of Crete. On 30 November, Papanikolis successfully attacked and sank a 6000-cargo vessel at the Alimnia islet near Rhodes. On January 17, 1943, after carrying agents and equipment to Hydra, she captured the 200-ton sailing vessel Agios Stefanos and manned her with part of her crew, which sailed her to Alexandria. The next day, she sank another 150-ton sailer. During subsequent patrols in March and May, she sank further four sailers. "Papanikolis" survived the war and returned to Greece after liberation in October 1944. The obsolete submarine was decommissioned from the fleet in 1945. The ship's conning tower was preserved and is on display in the Hellenic Maritime Museum at Piraeus. Subsequently, two other submarines of the Greek Navy were named "Papanikolis."

Grigel

Grigel – strong wind from the northeast, blowing off the coast of the island of Sicily, Italy, in the period from October–November to March. Its duration ranges from several hours to 5 days. With a continuous wind at a speed of 21–24 m/s, waves up to 6 m high can be observed. Grigel most often occurs after cyclones and is accompanied by poor visibility and heavy rain.

Gulf of Corinth

Gulf of Corinth (*Korinthou kolpos*) – the Gulf of the Ionian World off the western coast of Greece, between the mainland in the north and the Peloponnese in the south. The Gulf of Corinth is surrounded by the prefectures of Aetolia-Acarnania and Phocis in the north, Boeotia in the northeast, Attica in the east, Corinthia in the southeast and south, and Achaea in the southwest. Length – 130 km, width at the entrance – 3 km, the widest part – 35 km. It is situated in one of the most seismically hazardous areas in Europe. The southern coast of the bay is slightly indented. The eastern mountains in its middle part steeply descend to the sea. 20 km west of the Corinth Canal, the mountains turn into a plain covered with pastures and cultivated fields. Many rivers flow from this bank into the bay. On the slopes of the mountains in the valleys and along the railway from Patras to the Corinth Canal, there are towns and villages. The northern coast of the bay is indented. Crissaean Gulf and Bay of Antikyra, as well as several small bays form inlets into its shores. The area adjacent to this coast is a mountainous region with rocky peaks, the slopes of which in some places abruptly drop off to the sea. Mountain streams flowing through deep gorges flow into bays. There are several small islands near the coast, the largest of which are Trizonia Island and Alkyonides Islands. The mountains on the southern and northern shores of the Corinthian Gulf are covered with vast vegetation. On the lower slopes of the ridges, there are pastures and arable lands, and olive trees grow in the valleys. The gulf is deep – the depths in the middle part reach 900 m, and 65–70 m at the entrance. In the western part of the gulf, tidal currents are observed, the change of which occurs at the moments of full and low water at Trizonia Island. The tidal current flows into the gulf. In the east it is bounded by the Isthmus of Corinth, through which the channel of the same name was dug into the Aegean Sea, and in the west it is separated from the Gulf of Patras by the capes Rio and Antirrio, connected by the Rio–Antirrio Bridge. On the capes, there are Venetian castles preserved that guarded the entrance to the bay. They were built in 1686 during the invasion of the Venetians in Greece. (from 136) (Fig. 12).

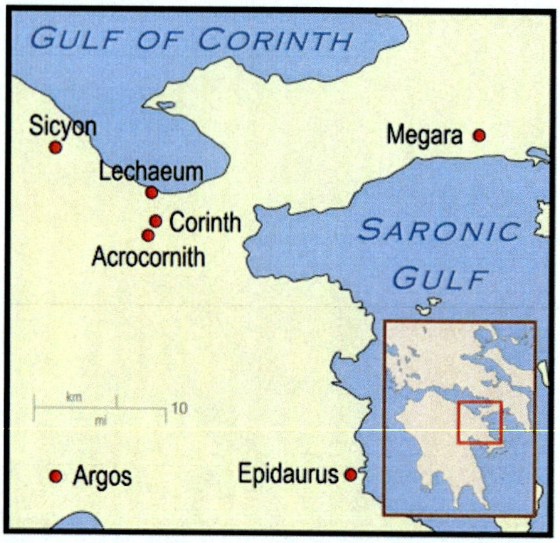

Fig. 12 Gulf of Corinth (Korinthou kolpos). (Source: https://upload.wikimedia.org/wikipedia/commons/thumb/e/ed/Isthmus_of_Corinth_in_ancient_Greece.svg/2000px-Isthmus_of_Corinth_in_ancient_Greece.svg.png)

Gulf of Kyparissia

Gulf of Kyparissia (*Kiparissiakos kolpos*) – it forms an inlet into the shore for 16.5 km into the western coast of the Peloponnese in the Ionian Sea, Greece, between the Cape Konello (Akra Kounellos) and Cape Katakolo. From Cape Konello the coast gradually lowers to the top of the bay, where it turns into a sandy beach, followed by a chain of shallow lakes for 30 km along the coast. Several small rivers flow into the bay, and the Alpheios River flows into its northern part. The shores of the bay are slightly indented.

The bay is open to westerly winds. The shores of the bay are densely populated. There are two large cities – Kyparissia and Pyrgos.

Gulf of Patras

Gulf of Patras (*Patraikos Kolpos*) – a deep gulf of the Ionian Sea. In fact, a vast strait leading to the Gulf of Corinth. It is situated between the north-western coast of the Peloponnese and the southern coast of the Aetolia-Acarnania nomos of mainland Greece. In the east, it is connected by a narrow strait with the Gulf of Corinth near the Rio-Antirrio Bridge. The length of the bay is 40–50 km, the width is 10–20 km, and the area of the water area is approximately 350 km^2. The average depth is 50–100 m. The entrance to the bay from the west is covered by the Ionian Islands. The northern shores of the bay are uneven. At the entrance to the bay from the west, the coast is low and swampy, but to the north and east it rises sharply. Along the eastern part of the northern coast of the gulf, there are high mountain ranges, which in some places come close to the coast. The coast has a rich and diverse Mediterranean vegetation. The mountain slopes are covered with bush and forest. Cultivated fields are located in the lowlands and valleys. The southern coast of the bay is slightly indented. Many rivers and mountain streams flow into the bay. The largest river in Greece, Acheloos, flows into the northern part of the deep-water area on the approach to the bay, and from the north, the second river of Greece, Evinos. The coast of the bay is densely populated. On the southeastern shore of the bay there is a large city and port of Patras, on the northern shore of the bay – the city and port of Missolonghi. Navigation is well developed in the bay. In 1571, one of the largest naval battles in history took place in the bay – the Battle of Lepanto. In 1772, during the First Archipelago Expedition of the Russian Navy, a battle took place in the bay between the Russian and Turkish squadrons, which ended in the victory of the Russian sailors.

H

Hellas

Hellas – the Greek name Hellas – Greece itself – and the Hellenes – the endonym of the Greeks – before, according to Homer's Iliad, was applied to the region in the southern part of Thessaly – Phthiotis. How this name became universal is unknown. Originally, the common name for all Greeks was the term "Panhellenes" ("all Greeks"). With the adoption of the term "Hellene," common to all Greeks, Hellas became a collective name for all of mainland Greece, and then all of Greece, including archipelagos, islands, and regions in Asia Minor (as opposed to the historical Magna Graecia located in southern Italy). At present, Hellas is the official self-name in Greece, and the words "Greek" or "Greece" are not recognized by the population and are used only in communication with foreigners. In other countries, Hellas is often synonymous with the concept of "Ancient Greece."

Hellenes

Hellenes – originally it was the name of a North Greek tribe. Then this name spread to all Greeks. From the fifth century BC, its application distinguished the Greeks from all foreign peoples (barbarians) who did not speak Greek. Greece was called the Kingdom of the Hellenes.

Hellenic Centre for Marine Research (HCMR)

Hellenic Centre for Marine Research (HCMR) – public research organization of the General Secretariat for Research and Technology (GSRT) of the Ministry of Education, Research, and Religious Affairs. The center includes the Institute of Marine Biology, Biotechnology and Aquaculture (IMBBC); the Institute of Marine Biological Resources and Inland Waters (IMBRIW); Institute of Oceanography; Branches: research and documentation support; technological advancement and product enhancement; and administrative and economic. The center has three research stations: Rhodes Hydrobiological Station, Chania Mediterranean Marine Station, and "Thalassocosmos" Cretaquarium. The center includes a fleet of three research vessels: "Aegaeo," "Philia," "Alkyon." The history of the creation of the HCMR dates back to 1912, when the Greek government decided to discuss the possibility of creating a Marine Research Institute as part of the research carried out by the Italian expert Professor D. Vinciguerra, Director of the Ichthyological Station in Rome. After discussing some of his proposals, the first Greek Maritime Institute, as the Marine Hydrobiological Station, was created in 1914 in Paleo Faliron. Soon after its creation in 1915,

I. S. Zonn, *The Ionian Sea Encyclopedia*, Encyclopedia of Seas,
https://doi.org/10.1007/978-3-031-08206-1

the first studies in marine biology and fisheries were carried out. In 1948, the station was renamed the Laboratory of Fisheries Studies and transferred to the Ministry of Agriculture. In 1945, after World War II, the Hellenic Hydrobiological Institute of the Academy of Athens in Piraeus was created in conjunction with the Hydrobiological Station in Rhodes, which was founded by the Italians during the occupation. At the same time, the small vessel Glauki was adapted as a research vessel, and in 1946 the first three Greek oceanographic expeditions were carried out. In 1965 the Hydrobiological Institute of the Academy of Athens was amalgamated with the Laboratory of Fisheries Studies, forming a new structure – the Institute of Oceanographic and Fisheries Research, which became fully functional in 1970. In 1985, as a result of new research legislation (law 1514), the Hellenic Center for Marine Research was created (Fig. 1).

Hellenic National Oceanographic Data Centre

Hellenic National Oceanographic Data Centre – it was created in 1986 within the framework of Greece's cooperation with the International Oceanographic Commission (IOC) of UNESCO. It is a national agency, part of an international network of National Oceanographic Data Centers operating under the International Oceanographic Data and Information Exchange Committee of the IOC. The objectives of the Center are the protection and effective management of oceanographic data and information, including accumulation, formatting, quality control, cataloging, archiving, and exchange of marine data and information. The Center operates under the Hellenic Center for Marine Research and manages oceanographic data collected by Greek public and private organizations,

Fig. 1 Aquarium, research centre and museum in Rhodes. This centre is currently administered by the National Centre of Marine Research. (Source: https://www.ayda.ru/images/places/10231/o_6594.jpg)

as well as data from exchanges and international oceanographic data centers through the center's work in European and international projects.

High-Body Pickarel, *Spicara smaris*

High-Body Pickarel, *Spicara smaris* – belongs to the Centrocanthidae family. This is a pelagic schooling bottom fish, with an elongated body, slightly flattened from the sides. The color is grayish with a yellow tinge on the sides, with blue longitudinal stripes of the living fish (which quickly disappear after being caught) and a dark spot in the middle. There are blue specks on the membrane between the rays of the dorsal and anal fins. The length usually does not exceed 20 cm, and it lives up to 4–7 years. During the lifetime of the high-body pickarel, a change of sex occurs – up to 2–4 years of juvenile fish age, the germ gland functions as ovaries, and at an older age as the testes. Therefore, the maximum length of males is 19.4 cm with a mass of 82 g, and females 15.7 cm and 45 g, respectively. When changing sex, the color also changes – from grayish females to bright males, who get mating outfit during spawning. Hermaphroditism is noted (the germ gland function both as ovaries and testes). The fertility varies from 6 to 63,000 eggs. It feeds on bottom invertebrates (mollusks, polychaetes, and crustaceans), algae, and partly small fish and their caviar. Meat of high-body pickarel is very delicious, used fresh and dried. It is recently used for the production of canned food in oil and tomato. The fish is an object of sport fishing.

Himara

Himara (*Himarë*) – a city in the southwest of Albania, on the Ionian Sea, opposite the northern tip of the island of Kerkyra. Included in the Vlorë county and region. Population – approx. 4000 people, mostly Greeks, but also Albanians. A typical resort town, the economy of which is developing on the basis of tourism. The city was founded by the Greeks during the Byzantine Empire on the territory of Northern Epirus, as evidenced by numerous excavations. Later it was part of the Albanian principalities of Muzakaj and Arianiti, and with the collapse of Byzantium and the Ottoman invasion of Albania, it passed to the Ottoman Empire under the control of the Kingdom of Naples until 1492. For some time the city was the site of the battles between the Ottoman Empire and the Venetian Republic for the right to control the entire region. By the seventeenth century the Greeks were the majority, but by the beginning of the eighteenth century the Greek population declined significantly due to the rule of the Albanian despot Ali Pasha of Tepelena. Until 1833, Himara constituted a separate Greek diocese, in fact representing a state within a state, the autonomy of which the Turkish government was forced to recognize. With the outbreak of the Balkan Wars on November 18, 1912, a major of the Greek gendarmerie, Himariot Spyros Spyromilios (q.v.) landed in Himar and liberated his homeland. On December 14, 1913, by a protocol signed in Florence, six European powers decided to include the entire Northern Epirus, including Himara, in the created Albanian state. On February 14, 1914, the protocol was brought to the attention of the Greek government. The Greek government, under pressure from Western governments and having received assurances that Greek control over the Aegean would be recognized, ordered the army to leave Northern Epirus. Spyromilios refused to leave Himara, planning to achieve reunification with Greece. The Himariots were the first of the Albanian Greeks to begin the struggle, proclaiming an autonomous state in 1914. Later, Himara became part of the "Autonomous Republic of Northern Epirus," proclaimed on February 28 in the city of Gjirokastër. Spyromilios, who left the Greek army, held the Himara-Tepelenë defense line throughout the entire period of the autonomist movement, but this did not bring the desired result – reunification with Greece. With the outbreak of World War I, Himara, like the entire Northern Epirus, became part of independent Albania. The old castle of the old town is rightfully considered as the main sightseeing of Himara. The churches of the old town (e.g., Panagi Kasopitra, according to legend, built on the place of the pagan temple of Apollo, and the All Saints' Church) and monasteries, towering on the mountain slopes, as well as the mansion of the Spyromilios family are wonderful monuments of architecture and history. To the south of Himara, on a peninsula connected to the coast by a thin strip of land, is an excellently preserved ancient building. From above, the fortress looks like a regular triangle. According to legend, in 1803 it was erected by Ali Pasha of Tepelena in honor of his wife Vassiliki (now it is abandoned). There is a church next to the fortress. The crystal clear waters of the Ionian Sea deservedly attract tourists from many countries of the world to Himara. The beaches are pebbly, but there are also several sandy ones. The cleanest Jale Beach is located near Himara (Fig. 2).

Homer

Homer (*Homeros*) – even in antiquity, nothing is known for certain about the life and personality of Homer. The chronological period in which modern science localizes the life of Homer is approximately the eighth century BC. According to Herodotus, Homer lived 400 years before him, which indicates a date of 850 BC. An unknown historian writes in his notes that Homer lived 622 years before Xerxes, which indicates 1102 BC. At the moment, there are several dates of birth and evidence for them. Homer's place of birth is unknown. In ancient tradition, seven cities argued for the right to be called his homeland: Smyrna, Chios, Colophon, Salamis, Rhodes, Argos, and Athens. The Greek historian Herodotus and Pausanias believed that Homer died on the island of Ios in the Cyclades

Fig. 2 City of Himarë. (Source: https://albaniapropertygroup.com/source/Himare-ALVS129/Himara%20-%201.jpg)

archipelago. Probably, the "Iliad" and "Odyssey" were created on the Asia Minor coast of Greece, inhabited by Ionian tribes, or on one of the adjacent islands. However, the Homeric dialect does not provide accurate information about the tribal affiliation of Homer, since it is a hexameter – a combination of the Ionian-Aeolian dialects of the ancient Greek language. Traditionally, Homer is portrayed as a blind man. Most likely, this idea does not come from the real facts of his life but from the reconstruction, which is a specificity of the genre of antique biography. There is a legend about the poetic contest of Homer with the first Greek, therefore, and European poet Hesiod, described in the essay "Contest of Homer and Hesiod," created no later than the third century BC, and according to many researchers, even much earlier. The poets allegedly met on the island of Euboea (or Evia) at the games in honor of the deceased Amphidemus and each recited their best poems. The victory was awarded to Hesiod, while the audience's sympathies were on Homer's side. In addition to the "Iliad" and the "Odyssey," Homer is credited with many epic poems created later: "Homeric Hymns" (seventh to fifth centuries BC, the most ancient examples of Greek poetry), the comic poem "Margites," "Cypria," etc. (Fig. 3).

"Husky"

"Husky" – a cryptonym for plan of the amphibious assault of Sicily by Anglo-American forces in 1943.

Hydrocarbons of the Ionian Sea

Hydrocarbons of the Ionian Sea – In 2014, Greece presented a program of concepts for oil and gas exploration in the sea area of Western Greece (the Ionian Sea) and south of the island of Crete (the Aegean Sea) in London. Thus, Greece is focusing on its own hydrocarbon resources in order to reduce the risks of excessive dependence on energy imports. About 20 marine areas (sections) with an area of 225,000 km^2 are allocated for exploration, which, as a result of hard and patient work in recent years, have been determined on the basis of 50 parameters, such as sea depth, geology and environmental conditions, etc. Later, a decision was signed to grant the rights to explore and use hydrocarbon resources in the Block-2 offshore area in the Ionian Sea. With the final approval of the contract with the consortium,

Fig. 3 Homer (*Homeros*). (Source: https://interesnyefakty.org/wp-content/uploads/gomer-3.jpg)

which includes Hellenic Petroleum, French Total, and Italian Edison, the way will open for exploration activities in the Ionian Sea for the first time. After the signing of an agreement, concerning sect. 2 west of Corfu, the signing of the contract for sections 1 (northeast of Corfu) and 10 (in the Bay of Cypress) is under consideration. The potential investor is Hellenic Petroleum, Greece. Contracts are also expected to be signed for three land plots in western Greece: Arta Preveza, Northwestern Peloponnese, and Energion (Aetolia-Acarnania).

I

Igoumenitsa

Igoumenitsa – a coastal port city located in the bay of the same name of the Kerkyra Strait, at the mouth of the Thyamis River, in northwestern Greece. The second largest port in western Greece. The population is 17,000 people (2006). Igoumenitsa is a city rebuilt after World War II. All historical and ancient buildings were destroyed during military bombing. Only a few old churches have survived. In ancient times it was a fishing village. During the period of the Turkish yoke in Greece, it was a small town called Grava. In 1913 it was liberated from the Turks, and after the creation of Thesprotia in 1938, it became its capital and took its present name. The mountains surrounding Igoumenitsa are bare and treeless (Fig. 1).

This is due to systematic fires and the burning of forests during the Turkish rule over Greece. In addition, the forests surrounding the city were badly damaged by a fire in 1990, however, then the forest cover was restored within 10 years. Over the past few years, Igoumenitsa has more than doubled its territory and expanded in all directions. The rapid growth was facilitated by the construction of a mega-freight terminal, which is currently the third largest in Greece after Piraeus and Patras, and the construction of the "Egnatia" highway, which runs through northern Greece and connects Igoumenitsa with Ioannina, Thessaloniki, Alexandroupolis, and Turkey. Igoumenitsa is a summer holiday center in western Greece. The main advantage of the city and its surroundings is mountain walks with views of wildlife and beaches. The main beach is Syvota, surrounded by olive groves, located 23 km from Igoumenitsa. Also popular is Drepano Beach, a 2 km long crescent-shaped sandy strip bordering the lagoon, and Sayiada, the last coastal village before the border with Albania. The swimming season lasts from May to November. Igoumenitsa is connected with the ports of Italy – Trieste, Venice, Ancona, Bari, Brindisi, Otranto, and the ports of the Ionian Islands.

Imbatto

Imbatto – the local name for the sea breeze, which usually sets at about 10 am. In the first 2–3 h it intensifies, becomes strongest by 3 pm, and then gradually weakens and stops after sunset.

Imeri Gramvousa

Imeri Gramvousa ("*Tame Gramvousa*") – an island, one of the two uninhabited islands of Gramvousa. Located northwest of the coast of Crete Island. The island hosts the remains of a Venetian fort, built in 1579 during the Venetian rule of Crete for the defense against the Ottoman Empire. The fort remained in Venetian hands throughout the Cretan War (1645–1669), which

I. S. Zonn, *The Ionian Sea Encyclopedia*, Encyclopedia of Seas,
https://doi.org/10.1007/978-3-031-08206-1

Fig. 1 The City Port Igoumenitsa. (Source: https://topgreekgyms.gr/wp-content/uploads/2019/02/Top_Greek_Gyms_Hgoumenitsa_Blog.jpg)

ended with the transfer of Crete to the Turks in 1669. Gramvousa, the Souda fortress, and the island-fortress of Spinalonga were retained by Venice. These fortresses defended the trade routes of Venice, as well as were strategic bases in case of a new war with the Turks, as well as the ruins of buildings erected here by Cretan rebels and pirates who alternately controlled the island during the Greek War of Independence of 1821–1830. The island is now a popular sightseeing in western Crete. The distance between the islands of Imeri Gramvousa and Agria Gramvousa is about a kilometer (Fig. 2).

Integrated Coastal Zone Management (*ICZM*)

Integrated Coastal Zone Management (ICZM) – an approach to economic development of coastal zones management without harming the environment. The European Commission considers ICZM as a permanent management process the overall goal of which is to introduce coastal preservation as well as maintaining their biodiversity into sustainable development practices. On the one hand, the ICZM is an economic and legal instrument for harmonizing the many conflicting interests of coastal users of natural resources and, on the other hand, it is a dynamic process of developing and executing coordinated strategies for the use of natural, social, cultural and institutional resources. The successful implementation of the ICZM should be supported by systematic information and scientific support, advanced engineering technology, and socioeconomic and legal mechanisms. The main instruments of the ICZM are: spatial planning of coastal and marine areas, intersectoral and interregional coordination, public involvement, state monitoring and assessment of changes, a comprehensive analysis of sustainable development indicators, and financial and legal mechanisms for the implementation.

Ionia

Ionia – city and port on the east coast of Sicily north of Catania, Italy. Winemaking and fishing are well developed here. Construction of small ships developed here. As well as the export of wine and citrus fruits.

Ionia Odos

Ionia Odos (*Ionian Way, Ionian Freeway*) – section of the highway, opened in 2018, along the

Fig. 2 Imeri Gramvousa ("*Tame Gramvousa*"). (Source: https://www.alargo.com/photos/contentpage/60/imeri-gramvousa-chania-crete-18-sep-2014-1-general-fullscreen.jpg)

coast of the Ionian Sea in western Greece. The construction of "Ionia Odos" was identified as one of the most difficult. The existing terrain has made it difficult to lay a highway that runs through the Pindus mountain range. The 196 km motorway connects the settlements on the Peloponnese, Amfilochia, and Arta and reaches Ioannina, where it connects with the "Egnatia Odos" motorway (670 km), opened in 2009. The construction gave a serious impetus to the development of the region, which is considered (according to EU data), the poorest in Europe.

Ionian Academy

Ionian Academy – the first institution of higher education on the island of Corfu (Kerkyra), which is considered the predecessor of the Ionian University. Founded by the English Lord Guildford. He set out to contribute to the intellectual development of the Greek nation and to open an institution of higher education in Ithaca for all Greeks. It was planned that the academy would consist of four schools: theology, law, medicine, and philosophy. From the outset of the academy's founding, Guildford faced a hostile mood that threatened the aristocratic monopoly "on higher education." His efforts have lasted for more than 8 years, the dream came true in 1811. Famous Greek scientists refused to teach at the academy and Guilford had to invite young scientists from Europe. Due to the outbreak of the Greek War of Independence in 1821, it had to be moved from Ithaca to Kerkyra. The official opening of the Academy in Kerkyra took place in 1824. Due to the death of Lord Guildford in 1828, the academy was closed, and a library was opened instead. In 1943, the library burned down after an air strike by the Italians. In 1944 the academy was reopened after reconstruction. After 20 years, it was closed after the unification of the Ionian Islands with Greece, as Greece could not afford to maintain two universities.

Ionian- Adriatic pipeline (*IAP*)

Ionian- Adriatic pipeline (*IAP*) – the 530 km gas pipeline planned for construction is, in fact, a branch from the Trans-Adriatic gas pipeline. It should link Fier, Albania, and Split, Croatia, through the territories of Montenegro and Bosnia and Herzegovina. Estimated cost – 610.5 million euros. In 2013, a Memorandum was signed between all the participating countries.

Ionian Archipelago

Ionian Archipelago (*Archipelago ionian*) – is a common name for a complex of seven islands in the Ionian Sea: Kerkyra, Ithaci, Zakynthos, Kefalonia, Lefkada, Paxi, Kythira. They are located to the west of the mainland – the Balkan Peninsula and belong to Greece.

"Ionian cruises"

"Ionian cruises" – Greek maritime company serving the Ionian Islands, as well as making exclusive tours to Albania. Based on the island of Kerkyra (Corfu). The company has five vessels: two hydrofoil dolphin vessels – Skendrben and Santa III – designed for 124 passengers, speed 33 knots; three large passenger boats – "Antigoni," "Christina," "Sotirakis I," designed for 250 people, speed 15–17 knots.

Ionian Islands

Ionian Islands – include a group of seven large islands: Paxi, Zakynthos, Kefalonia, Lefkada, Ithaca (or Ithaki), Kerkyra (Corfu), Kythira, as well as the adjacent islets and cliffs located along the Balkan western coast of Greece and in the south along the western part of the Peloponnese for 260 km. Between these islands and the western coast of Greece there is a deep-sea basin, from the middle of which the Gulf of Patraikos begins, passing further east into the Gulf of Corinth. They are sometimes referred to as the "Heptanese (Seven Islands)." In fact, there are many more islands in this system – twelve. Due to their geographical position – they are at the crossroads of three continents – these islands have been inhabited since ancient times and have developed great civilizations. All the Ionian Islands impress with their beauty. The cities of these islands are built in the style of Venetian architecture. Everything here is buried in greenery, and olive groves and pine forests descend to the very water of the Ionian Sea. Numerous, several-kilometers long as well as mini-beaches with white sand and azure sea are marked with the Blue Flag. They rate popularity among tourists from all over the world. Mountain villages and picturesque harbors with moored yachts and sailboats, paths and manicured paved roads, waterfalls and mountains, water areas, and marine parks where turtles and seals live are situated here. The importance of the Ionian Islands was emphasized by the French emperor Napoleon: "The islands of Corfu, Zante and Kefalonia are more important to us than the whole of Italy." The Ionian Islands have a very rich history, as evidenced by a large number of medieval fortresses and monasteries, as well as mythical buildings and museums (Fig. 3).

The Ionian Islands have been inhabited since the Neolithic times, as evidenced by archaeological finds on the island of Lefkada. However, the intensive colonization and development of the islands falls on 1200–1400 BC accounted for the heyday of the Mycenaean civilization. Approximately these times are described by Homer in the poems "Iliad" and "Odyssey." The poet located the kingdom of Odysseus on one of the seven main islands of the archipelago Ithaca. In those days, the islands were considered a remote, insignificant province of Hellas. However, in the eighth to seventh centuries BC, the islands of Corfu and Lefkada were captured by the Corinthians, who needed a foothold for the conquest of Sicily and further advance to the west of the Mediterranean Sea.

Since that time, the Ionian Islands have been incessantly passed from hand to hand, the reason for which was their strategic importance as a key to the entire eastern Mediterranean. Any invader felt safe behind the walls of the impregnable island fortresses; however, he himself could

Fig. 3 The Ionian Islands. (Source: https://simpsontravel.blob.core.windows.net/media/26414/new-greece-main-map.jpg)

threaten all the surrounding lands from here. However, at all times the islands remained Greek in spirit. Thus, arts flourished here, and outstanding writers, painters, and musicians worked here. In 431 BC the struggle for the Ionian Islands became the pretext for the Peloponnesian War between democratic Athens and oligarchic Sparta. In the fourth century BC the islands became part of the Macedonian empire, in 146 BC passed under the rule of the Romans. For the next several centuries, the islands were the object of the constant struggle between the Byzantines, Saracens, Normans, Sicilians, and Neapolitans. In 1204, the Venetians occupied most of the islands and held them for several centuries. The Venetians managed to keep the islands from the claims of the Ottoman Turks, and while the whole of Greece was suffering under Turkish rule, the islands enjoyed relative autonomy. Each of the conquerors altered the names of the islands in their own way. So, under the Venetians, Kerkyra began to be called Corfu, Ithaca – Val di Compare, Kithira – Cirigo, Lefkada – Santa Maura, Zakynthos – Zante. In the eighteenth century, the islands become a haven for all the fighters for Greek independence. In 1797, during the Napoleonic Wars, the Venetians left the islands and the French settled here. However, already in 1799, the Russian-Turkish fleet under the command of the Russian naval commander F. Ushakov expelled the French from the islands. With his participation, in 1800 an agreement was signed on the formation of the Septinsular Republic (Heptanēsos) under the protectorate of Russia and Turkey, which existed for 7 years. In 1815, the independent State of the Ionian Islands was created here under the control of England. Only in 1864 did the British leave the islands, which became a province of Greece. During the Second World War, the islands were occupied first by the Italians, then by the Germans, until 1944. The earthquake of 1953 was the most significant event of the post-war history, which completely destroyed the villages and cities of Kefalonia and Zakynthos. Today this archipelago is a separate administrative district of Greece. The Ionian Islands are the only region of Greece that has never been ruled by the Ottoman Empire.

Ionian Islands School

Ionian Islands School (Ionian artists school) – a post-Renaissance regional art movement inspired by the Zakynthos painter Panagiotis Doxoras. This style arose as a result of the influx of painters from Crete who fled from the Turkish yoke. The collection of works of artists of this school is located in the Byzantine Museum or the Zakynthos Museum in Zakynthos.

Ionian riviera

Ionian riviera – part of the attractive beaches of the Ionian and Adriatic Seas, stretched along the national highway Sarandë – Vlorë, Albania. The locals call the Riviera. Bregu – the coast. The main sightseeing is Palasë the beach of a white sand, which is called "the white road."

Ionian Sea

Ionian Sea (*Lat. Mare Ionium, Greek. Ionic pelagos, Ital. Mare Ionio*) – is located in the central part of the Mediterranean basin, located between the Balkan and Apennine peninsulas and the islands of Crete and Sicily. The sea was praised by Homer and found a place in the annals of the ancient Romans. It is the location of many ancient Greek myths. Through the Strait of Otranto it connects with the Adriatic Sea, and through the Strait of Messina – with the Tyrrhenian Sea. It has no strict southern border. It washes the shores of southern Italy (regions of Sicily, Calabria, Basilicata, and Apulia), Albania (Vlorë Prefecture), and Greece (administrative districts of the Ionian Islands, Epirus, Western Greece, Central Greece, Attica, Peloponnese, and Crete). The International Hydrographic Organization defines the following boundaries of the Ionian Sea. In the north, the border follows from the mouth of the Butrint River in Albania to Cape Karagol on the island. Kerkyra, then along the northern coast of the island of Kerkyra to Cape Kefali and from here to Cape Santa Maria di Leuca in Italy. In the east from the mouth of the Butrint River in Albania down the coast of the mainland to Cape Matapan. In the south from Cape Tainaron to Cape Passero – the southern point of the Sicily Island. In the west along the eastern coast of the island of Sicily and the south-eastern coast of Italy to Cape Santa Maria di Leuca. The island of Kythira is where the Ionian Sea meets the Aegean Sea. In ancient times, the life of Ancient Greece was associated with the Ionian Sea, especially the Ionian Islands belonging to the ancient Greeks. For the Greeks, the Ionian Sea meant primarily the present Adriatic Sea. According to one of the widespread versions, the name of the sea dates back to the history of the ancient region of Ionia, which was formerly located near modern Izmir in Turkey. Traditionally, the Ionians are among the four main ancient Greek tribes. In the eighth to sixth centuries BC these people managed to settle on the island of Kefalonia from the system of the Ionian Islands, and began to develop the adjacent waters. Another version links the name of the Ionian Sea with the name of the heroine of ancient Greek myths Io – the daughter of the river god Inachus (Fig. 4).

According to legend, this girl was a priestess of Hera and the beloved of Zeus the Thunderer. To hide his passion from the face of his jealous wife, Zeus turned Io into a cow, but Hera saw through his trick and asked for a snow-white animal as a gift. Zeus could not refuse, and Hera ordered the hundred-eyed Argus to guard Io. When Hera found out that Hermes had killed Argus, she sent a terrible gadfly to Io, which stung the poor animal and drove it around various lands, among other places passed Io and the sea, which is now called the Ionian. Homer's poem "The Odyssey" takes place on Ithaca, one of the Ionian Islands. Hence the sea is sometimes called the Odysseus. According to Homer, Ithaca is the homeland and kingdom of Odysseus, the place from where Greek ships sailed to Troy. There were also events in the history of the Ionian Sea, the historical authenticity of which is beyond doubt among researchers. On September 2, 31 BC near Cape Actium (63 BC–14 AD) the last naval battle of

Fig. 4 Ionian Sea. (Source: https://www.captaincork.com/wp-content/uploads/2016/04/Ionisches-Meer.png)

antiquity took place between the fleets of Gaius Julius Caesar, on the one side, and Mark Antony and Cleopatra, on the other. About 620 ships of both fleets entered the Ionian Sea. They fought for power over the Roman Empire. Unfortunately for Mark Antony, his ships were not so maneuverable, and therefore they turned out to be an easy target for the lighter ships of Octavian Augustus. Cleopatra, realizing that luck was not on her side, simply withdrew from the battle with her fleet, and Antony, having lost hope of reinforcements, followed her. As a result, Caesar was victorious, and a monarchy was actually established in the state: the new emperor reigned supreme over the Roman Empire. The Ionian Sea is the deepest part of the Mediterranean. The western coast of Greece is the southern and western coast of the Peloponnese and the southern part of the western coast of the Balkan Peninsula. These mountainous peninsulas are separated from each other by the Patraikos and Corinthian bays deeply protruding into the mainland. They are connected only by a narrow isthmus through which the Corinth Canal was dug – the shortest route between the Aegean and Ionian Seas. The western coast of Greece is predominantly high and steep. The most extensive lowlands, marshy in places, open to the Gulf of Patraikos and to the northern shore of the Gulf of Amvrakikos. Some landmarks can be identified from 110–150 km. The Ionian Islands, which are sometimes called the "seven islands" by the number of the largest land areas, stretch along the Balkan coast of the sea: Kerkyra (Corfu), Paxos, Lefkada, Ithaca, Kefalonia, Zakynthos, and Kythera. The mountain ranges here, as well as on the Balkan Peninsula, have a mainly meridional direction. The height of the mountains is 800–1000 m. Only individual peaks on the island of Kefalonia – the largest of the Ionian Islands – rise on more than 1500 m. In fact, there are many more islands in this system. The area of the Ionian Sea is 169 thousand km^2, the maximum depth is 5121 m, which is the greatest depth of the Mediterranean Sea. Near the coast, the bottom does not go deep; on many beaches near the coast, shallow water is noted. The bottom has the shape of a hollow, covered with deposits – mainly silt, and closer to the shores – silty sand, sand, and shell rock. Pumice is found in areas of volcanic activity. The shores of the Ionian Sea are highly dissected, especially in the eastern part, where the Ionian Islands are located. Gulfs of the Ionian Sea are: Patraikos, Corinthian (the deepest in the Ionian Sea), Taranto, Mesiniakos, Amvrakikos, Kyparissiakos, etc. The range of water temperature varies is from 14 °C in February to 25 °C in August. At the bottom, the water temperature is about 13 °C. Salinity exceeds 38‰, and it increases from west to east – this is the saltiest sea in Greece. The water has a deep blue tint. It is sometimes called violet because it often takes on a violet hue at sunset. The tides are semi-diurnal (up to 0.4 m). The sea water is quite calm. The main flux of the constant current of the Ionian Sea comes from the Atlantic Ocean, through the Strait of Gibraltar, along the coast of Africa, generally from west to east. The average speed of a constant current in most of the sea is mostly less than 0.5 knots, in some places – 0.6–1 knots. In conditions of sustainable and strong winds, the direction and speed of the constant current change noticeably, in some cases the direction changes by 180°. Surface currents have a speed of about 1 km/h and form a cyclonic circulation. The Ionian Sea is deep. The 100 m isobath passes at a distance of no more than 10 km from the coast, and in some places – very close to it. At a distance of more than 10 km from the coast, the depths sharply increase to 1500 m, and further, the bottom topography becomes relatively flat. Some depressions have depths of 3500 m and more. The maximum depth in the Ionian Sea is 5121 m, which is located 124 km west of Cape Tenaron. In the narrowest part of the Isthmus of Corinth, which connects the southern part of the Balkan Peninsula with the Peloponnese, the Corinth Canal was dug, which reduced the sea route from the Ionian Sea to the Aegean by about 275 km. The Strait of Messina, separating the island of Sicily from the southern tip of the Apennine Peninsula, stretches for 550 km from north to south. This strait is used by ships sailing from the Ionian Sea to the Tyrrhenian. The Ionian Sea is located in the subtropical zone, the climate of which is characterized by dry, hot summers (May–September). The air temperature is relatively high. Its average annual value is almost everywhere 19–22°. The diurnal variation in air temperature is quite noticeable and most manifested from May to September, when the daily amplitude is 7–10°. Relative air humidity averages 50–80%. In the open sea, winds from the northwest and west prevail throughout the year, in addition, winds from the north and southwest are frequent. The wind speed from November to April is higher than from May to October. From May to October, it is 2–5 m/s. Storms are rare in this area. Sometimes storms are violent, even turning into hurricanes, and are accompanied by downfalls. There are frequent squalls, accompanied by downfalls and hail, in which visibility is significantly impaired. A wind similar to bora, however, weaker than it, is known locally as borino. It is observed from May to September. Sirocco – hot wind from the south and southeast during the warm period of the year and moderately warm during the cold period – in the western part of the sea, it is observed almost during the whole year. Etesian – steady wind of northern directions – is usually observed from mid-May to mid-September in the area of the western coast of Greece. Taranta – strong wind from the northwest, which can last continuously even for a day, from May–June to September. Mists are extremely rare. Haze is more common than mist. In a significant part of the sea, visibility of approximately 20 km and more prevails throughout the year, its frequency is 60–85%. Winds have a significant impact on visibility. Thus, in conditions of sirocco, visibility drops sharply (sometimes up to 1 km or less), and in conditions of bora, on the contrary, it increases (up to 20 km or more). Average monthly cloud cover varies mainly from 4 to 6 points. In September, the frequency of occurrence of a clear sky increases to 55–60%, and that of a cloudy sky decreases to 20–25%. Annual precipitation is 500–800 mm. The exceptions are the islands of Kerkyra and Zakynthos, located near the western coast of Greece, where their average annual number is 1110–1200 mm. The period October–April is rainy, and May–September is dry. During the rainy period, the maximum precipitation occurs in November and December, when the average monthly amount of them on the islands of Kerkyra and Zakynthos reaches

185–234 mm, on the western coast of Greece and on the island of Sicily – 100–142 mm, on the southeastern coast of Italy – 65–70 mm. The maximum amount of precipitation per day in these months can reach 115–175 m. In the dry period, the minimum precipitation is observed in June, July, and August, when on the islands of Kerkyra and Zakynthos, on the western coast of Greece and on the island of Sicily, an average precipitation per month reaches 25 mm, and on the southeastern coast of Italy, where precipitations reach 10–20 mm. The average annual number of days with precipitation varies mainly from 56 to 87, reaching 101–117 only in some places. During the rainy period, the average monthly number of days with precipitations varies from 5 to 14 (in some places from 10 to 16), and in the dry period – from 1 to 7. Precipitation is usually heavy. In some cases, they are very intense and can cause flooding. Snowfalls are very rare – only from December to February. The hydrological regime of the sea is characterized by a predominance of wave heights of 1–2 m, rather high temperatures, significant salinity, and water density. Seasonal sea level fluctuations are caused by increases or decreases in precipitation, evaporation, and coastal runoff. The tides in the Ionian Sea are predominantly semidiurnal, and only at certain points of the island of Sicily are irregular semidiurnal. The average tide exceeds 0.2 m, and the average syzygy tide is 0.3 m. Wave heights of 1–2 m prevail throughout the year. Their frequency is 40–50%. From March to November, wave heights of less than 1–2% are also frequent. Hydrometeorological conditions for navigation of ships in the Ionian Sea and in the area of Sicily island are most favorable from May to September, when clear weather prevails with weak winds and low precipitation, and strong waves are rare. Difficulties for navigation can arise with the "sirocco" wind, and in some areas – with the local winds "bora," "etesian," "taranta," "tormenta," and "gregale." Navigational safety is ensured by a sufficient number of aids to navigation. At the most important points, beacons with a visibility range of more than 35 km are installed. Some lighthouses have radio beacons. Harbor entrances and port basins are usually indicated by lights. Economic activity: fishing (mackerel, tuna, flounder, mullet); resorts, main ports: in Greece – Patras, Kerkyra, Igoumenitsa, Kefalonia, Ithaca, Lefkada, in Italy – Taranto, Catania, Nice, Rocca Imperiala, in Albania – Sarandë. Presently, the coast of the Ionian Sea is densely populated. The warm waters of the Ionian Sea attract tourists from all over the world. Tourism is flourishing almost all over the coast (Fig. 5).

The tourist season at the resorts begins from May to June and lasts till October. There are numerous tourist cities built on the place of

Fig. 5 Ionian Sea. (Source: https://www.grekomania.ru/images/greek-articles/other/big/288_moryagrecii-7.jpg)

I

ancient settlements, thus there can be found historical monuments in their vicinity – various military structures of antiquity, medieval castles, etc. For most of their inhabitants, the Ionian Sea is a source of income: someone gains profit from the services of tourists, someone from fishing.

"Ionian"

"Ionian" – a book by famous yachtsman Rod Heikell from New Zealand.

Ionians

Ionians – one of the three main Greek tribal groups, the most homogeneous in territorial, linguistic and cultural relations. The Ionians were believed to be the first to migrate to Greece (Attica, Euboea, Phocaea, Miletus) and inhabited the ancient Greek region of Ionia, which, according to legend, arose at the end of the second millennium BC. In ancient Asia, the Greeks were called "derivatives" (from the word "Ionian"), and they are considered the founders on the local shores.

Ionic Republic – United States of the Ionian Islands

Ionic Republic – United States of the Ionian Islands – the historical state that existed from 1815 to 1864. The Ionian Islands for centuries were part of the Venetian Republic, until it was transferred to revolutionary France according to conditions of the Treaty of Campo Formio, 1797). In 1798–1799. Russian-Turkish forces drove the French from the islands and formed the Septinsular Republic, which existed from 1800 to 1807. After that, France annexed the Ionian Islands again, and they became part of its Illyrian provinces. In October 1809, the British defeated the French fleet at Zakynthos and captured the islands of Zakynthos, Kefalonia, and Kythira. In 1810, the British occupied Lefkada island. Corfu (Kerkyra) remained under French rule until 1814. The Congress of Vienna authorized the transfer of the islands to the "amical protectorate" of Great Britain, which established a military administration there, and the Austrian Empire – a commercial status. In August 1817, the "Maitland Constitution" was ratified, according to which a federation of the seven islands was formed, and Sir Thomas became its first Lord High Commissioner. On March 29, 1864, Great Britain, Greece, France, and Russia signed an agreement in London, after the ratification of which the islands were to be transferred to Greece. On May 28, the High Commissioner issued a proclamation on the unification of the Ionian Republic with the Kingdom of Greece.

Ionic shores

Ionic shores – belong in the west of Italy, in the east – Greece and Albania. The western coast of Greece is the southern and western shores of the Peloponnese and the southern part of the western Balkan Peninsula from Cape Tainaron to the Gulf of Patras, stretches for 247 km. These mountainous peninsulas are separated from each other by the Patras and Corinthian (Greece) gulfs running deep inland. They are connected only by a narrow isthmus through which the Corinth Canal was dug – the shortest route between the Aegean and the Ionian Seas. The western coast of Greece is predominantly high and precipitous. The southeastern coast of Italy is higher and bolder than the western coast of Greece. The southern spurs of the Apennine Mountains stretch along it, several peaks of which, located only 15–20 km from the coastline, reach 2000 m or more. The spurs of the coastal mountains steeply cliffing into the sea, forming significant capes. In places where the direction of the mountain ranges coincides with the direction of the coastline, the coast is high and weakly indented. Between Cape Akritas and the Gulf of Patras), the coast gradually goes down and turns into a plain with rare hills. The mountains rise far from the sea, and behind them there are the mountain ridges, specific for the Peloponnese. In this area, the coast is strongly indented – the vast gulfs of Mesiniakos and Kyparissia running inland, as well as several bays (Navarino and others). From the western coast of the Peloponnese peninsula, many rivers and streams flow into the sea, most of which dry up in summer and significantly spill in winter. The largest river is Pamisos River, which flows into the Gulf of Mesiniakos. The rivers carry a lot of sand and silt, which are deposited at their estuaries and form bars. The coast is covered with rich vegetation typical for the Mediterranean region. On the lower slopes of the mountains, in the valleys, and on the plains, there are fields with crops of wheat and corn, vineyards and plantations of olive trees. The slopes of the mountains up to a height of 200 m are covered with maquis, in the higher areas there are oak forests, then they are replaced by beech massifs at an altitude of more than 1000 m – coniferous trees. There are many settlements on the western coast, the largest of which are Kalame, Kyparissia, and Pyrgos.

The main occupations of the population are the production of olive oil, winemaking as well as agriculture, and fishing. The beaches are especially good here. From the Gulf of Amvrakikos (Ambracian Gulf) to the southern entrance to the Kerkyra Strait, the coast extends 65 km to the northwest. It is mostly high and rocky, and only in places low, bordered by sandy beaches. A range of mountains stretches along the coast, the most significant of which is Mount Zalongo with a height of 849 m. Several rivers and mountain streams flow from the coast, the most significant river is Acheron, flowing along a broad, in places marshy plain. Due to different exposure, these shores develop under the impact of various wave systems. For the Greek coast in the cold season, the wave-forming winds are from the south-west, and in the warm season – from the north-west and north. Weak and moderate waves prevail, reaching the greatest strength from November to April. The southern coasts of Italy are heavily influenced by waves from the south and southeast during most of the year. Of the local winds, the bora, sometimes reaching the strength of a storm, has a certain value. The tides in the Ionian Sea are extremely low (0.1–0.2 m) and only in the Gulf of Patras reach 0.7 m. A more significant rise in sea level is usually associated with the action of southern surge winds. Under the impact of these winds, in the Strait of Corfu, the possible rise of the sea level reaches 1.2 m. The shores of the Ionian Sea are predominantly mountainous and are characterized by a significant indentedness. This is especially specific for the coast of Greece, which is accompanied by many mountainous islands (Fig. 6).

Here, steep limestone slopes of the Pindus mountain ranges up to 1000–2000 m high approach the sea. For most of this coastline, alternating rocky cliffs and various bays and gulfs with sandy and sandy-pebble beaches at their tops are typical. Due to the prevalence of weak waves, the abrasive sections of the coast change slowly. Capes usually have narrow benches, often with a corrosive surface. On limestone shores, the main development process is chemical abrasion, which is associated with the forming of leaching niches at the sea's edge. According to the available observations, the process of such niches forming is significantly accelerated near the exudation of fresh groundwater from crumbling rock, which has a high potential for lime dissolution. Rivers are the main source of sediment supply to the coastal zone. For example, 1800 thsd. m^3 of sand is annually supplied to the Gulf of Patras from rivers, and only 1000 m^3 from the abrasion of cliffs. Therefore, accumulative areas of the coast in most cases are located near river deltas. At the top of the Gulf of Messiniakos, due to processing by the outflow waves of the Pamisos river and neighboring small rivers formed in the late Holocene a sandy-gravel coastal plain (width 0.4–0.5 km), adjacent to the dead cliff. The coastal plain consists of a series of coastal ramparts that mark the gradual build-up of this area. The top of the neighboring Laconicos Gulf has a similar appearance. The small (10 km long, 4 km wide) erosional-tectonic Bay of Navarino, located in the southwestern corner of the Peloponnese and separated from the Ionian Sea by a narrow limestone ridge, has been actively filled with alluvium over the past 9000 years. By the beginning of our era, a large barrier complex with the isolated Osmanaga

Fig. 6 Ionian shores. (Source: https://wallbox.ru/resize/2560x1600/wallpapers/main2/201721/bereg-okean-u-okeana.jpg)

lagoon had formed at its summit. From the side of the Ionian Sea, a wide sandy beach develops here, bordered by migrating fields of dunes up to 10 m high. Lagoon-barrier complexes are also formed in the northern part of the wide Gulf of Kyparissia and south of the Gulf of Patras. The complex of coastal plains on the northwestern coast of Gulf of Patras is formed by adjoining deltas of two rivers, the main of which is the Aspropotamos River. In addition to the modern channel, which has been functioning for at least 2300 years, three more ancient channels of this river are clearly different in the topography, the age of which increases from west to east. The volumes of the river's solid runoff (before regulation in the 1960s – 3–4 million tons of suspended and two million tons of transported material per year) were quite sufficient for the forming of a vast delta plain by the successive development of four river aspirails. The dying off of the branches of the river occurred due to the clogging of the mouth with sand, which was facilitated by the active impact of sea waves. The estuarine areas of the dead branches are quickly eroded by the sea (Fig. 7).

The sand is processed by waves into bay-bars and barrier beaches, behind which shallow lagoons occur. The Italian shores of the Ionian Sea are also predominantly mountainous. Along the coast of the Calabria Peninsula, the mountain ranges of the Calabrian Apennines stretch, forming steep coastal slopes. Within the Gulf of Taranto, the mountain ranges recede somewhat from the sea. The shores are low-lying, bordered by beaches. The main movement of sediments is directed from south to north. In the same direction occurs a change in the composition of the beach material from gravel to sand-gravel and further, north of the mouth of the Agri river – to the sandy. On the side of the coast, the beaches are often bordered by dunes, sometimes up to 18 m high. In the area of the mouths of the main rivers of the region (Bradano, Agri, Sinni), the beach adjoins the edge of a marshy alluvial plain. A comparison of successive aerial photographs for 1943–1974 revealed a significant erosion of the shores of the Gulf of Taranto and a steady retreat of the coastline over the past 30 years at a rate of up to 4–6 m/year. The reason for the intensification of

Fig. 7 Ionian shores. (Source: https://kipmu.ru/wp-content/uploads/ionicheskoe.jpg)

erosion by the strongest southeastern waves is an increase in the deficit of sediments in the coastal zone. This is due to the limitation of alongshore sediment movement by artificial dams and the removal of huge volumes of sand and pebble material from beaches and riverbeds for building activity. Only for 1965–1977 35 million m^3 of alluvial material was extracted from the main rivers of the Gulf of Taranto, which significantly reduced the number of river flows into the coastal zone. The shores of the Ionian Sea are densely populated. There are numerous tourist towns here.

Isola delle Correnti

Isola delle Correnti (*Isola-della-corenti*) – "Island of streams" is the southernmost point of Sicily Island, Italy, where the Ionian and Mediterranean seas meet, creating foamy waves. This attracts surfers and windsurfers here. The area is considered one of the windiest in this area. In good weather, you can clearly observe the border of the seas, differing in color. The island is characterized by a long and beautiful golden sandy beach that stretches east and west. The beach is considered one of the best in Sicily. Nearby is the ancient Spanish Tafuri castle, where a luxury hotel is now located. It is known that the famous Formula 1 racer M. Schumacher (Germany) once wanted to acquire the castle. In addition, the beach got its name from a small island connected to Sicily by a small strip of stones, which, depending on the tide, is often submerged in water. In clear sunny weather it can be easily crossed. The island is home to a former tuna factory – Tonnara di Porto Palo. In November 1987, the island was included in the list of parks and reserves. Various species of animals live on its territory, and during the period of migration from northern Africa, various species of gulls, such as albatrosses, find shelter here (Fig. 8).

I

Fig. 8 Isola delle Correnti (*Isola-della-corenti*). (Source: https://i.ytimg.com/vi/OL61TZSLgi8/maxresdefault.jpg)

Isthmus of Corinth

Isthmus of Corinth – connects the Peloponnese peninsula with the Balkan Peninsula – the mainland of Greece between the Corinthian and Saronic Gulfs. It is situated near the city of Corinth, which gave the isthmus its name. Its length is approximately 42 km, width at the narrowest point is approximately 6 km. At the end of the XIX century at the narrowest point of the isthmus, the Corinth Canal was built. However, even before the existence of the canal, the Isthmus of Corinth from ancient times served not only to connect the Peloponnese with the rest of Greece, but also to shorten the waterways from the Saronic Gulf to the Corinthian Gulf. The Athenian historian Thucydides (460–396 BC) has a description of how the isthmus was used as a drag, called "Diolkos."

Italian Navy

Italian Navy – an independent branch of the armed forces of Italy. It consists of the Navy, Navy Air Force, and Marine Corps. In peacetime, they are under national subordination and solve the following main tasks: protecting sea communications and borders, ensuring a favorable regime in the zone of responsibility, conducting reconnaissance over the activities of ships and vessels of foreign navies in cooperation with the fleets of allies. In wartime, it is envisaged to transfer most of the Italian navy to the command of the allied naval forces of NATO at the South European Theatre of Military Operations (TMO). They will be recruited to support the combat operations of the strike naval forces to conquer dominance in the central region of the Mediterranean Sea (including the Tyrrhenian, Adriatic, and Ionian Seas), protect sea lanes, search and eliminate enemy submarines on anti-submarine lines, participate in amphibious operations, and ensure uninterrupted functioning systems of material and technical supply, support of ground forces in coastal areas. The general leadership of the Navy is entrusted to the Chief of the General Staff of the Armed Forces, direct – to the Chief of the General Staff of the Navy, who actually performs the functions of the commander. Organizationally, the Navy includes the fleet, aviation squadrons, the forces of four naval districts and two island naval commands, the command of the combat

Fig. 9 Italian Navy. (Source: https://ic.pics.livejournal.com/bmpd/38024980/6872787/6872787_original.jpg)

swimmers, and saboteurs "Teseo Tesei." The combat strength of the fleet includes three divisions of surface ships and three brigades (submarines, corvettes, and mine-sweeping forces). The Commander of the Fleet (headquarters in Taranto) is also the Commander of the Allied Naval Forces of NATO in the central region of the Mediterranean. The Ionian Sea is part of one of the four naval districts – the Ionian (Taranto) district. The Marines are represented by the "San Marco" Marine Brigade (four companies). Based in Brindisi and part of the third Fleet Division (Fig. 9).

Italian Peninsula

Italian Peninsula (*Penisola Italiana*) – see Apennine Peninsula.

Italy, The Italian Republic

Italy, The Italian Republic (Italia, Repubblica Italiana) – is a state in southern Europe, in the center of the Mediterranean. The country is named after the Italiki tribe ethnonym. Sometimes "Italy" is called "Apennine boot," as the peninsula resembles the shape of a shoe. It borders with France (the length of the border is 488 km) in the north-west, with Switzerland (740 km) and Austria (430 km) in the north, and Slovenia (232 km) in the north-east. It also has internal borders with the Vatican (3.2 km) and San Marino (39 km). It occupies the Apennine peninsula, Padan plain, the southern slopes of the Alps, the islands of Sicily, Sardinia, and a number of small islands. The area of the territory is 301.3 thsd. km^2.

Italy is predominantly mountainous country, where 77% of the territory is occupied by hills. In the north, there are located the southern slopes of the Alps with the highest point of Western Europe, Mont Blanc (4810 m), in the south from the Alps there is Padan plain. The Apennine mountains are stretched along the peninsula (the highest point is the mountain Corno Grande, 2914 m). The Apennines are divided into: Ligurian, Toscano-Emilian, Umbro-Marka, Abruzzi, Campanian, Lucani, Calabrian, and Sabini mountains. In the eastern part of the peninsula, there is the Gargano peninsula, in the south-east and south-west – the Salentin and Calabria peninsulas, respectively. There are active volcanoes – Vesuvius, Etna,

Stromboli. Earthquakes are quite frequently occurred. Italy is washed by the warm Mediterranean Sea and its component parts: the Ligurian, Tyrrhenian, Adriatic, and the Ionian Seas. From the east, the Apennine Peninsula is washed by the Adriatic Sea with the Gulf of Venice in its northern part. The strait of Otranto between Apulia and Albania connects the Adriatic Sea with the Ionian Sea. Between Apulia and Calabria, the Gulf of Taranto cuts deep inland. The very narrow Messina Strait separates Calabria from Sicily, and the Sicilian (or Tunisian) Strait, 135 km wide, divides Sicily from North Africa. To the north of Corsica there is the Ligurian Sea with the Gulf of Genoa. The length of the sea borders exceeds 7450 km.

Most of the small islands are divided into archipelagoes, such as the Tuscan archipelago, which includes the island of Elba, to which Napoleon Bonaparte was exiled (Fig. 10).

Only in the north there are large rivers. The largest of them are located in the Alps. The longest river in Italy is Po River which is 682 km long. Other significant rivers are the Adige (410 km), Adda (313 km) in the north of the country, and the Tiber (405 km) in its central part. Italy is rich in lakes, the largest lake is Garda (370 km^2), Lago Maggiore (212 km^2), Como (146 km^2), etc.

There is a large variety of minerals in Italy. However, many of these fields are poor, scattered throughout the country, and often occur uncomfortable for development. Thus, in 1982, iron ore mining was completely stopped in the country, including on the Elba Island, where the Etruscans mined it.

Lead and zinc ores with an admixture of silver and other metals are more abundant in Italy. These deposits are located on the Sardinia Island and the Eastern Alps. The region of Tuscany is rich in pyrite and mercury ore – cinnabar. Italy takes the second place in terms of cinnabar deposits. Antimony ores are deposited in the limestones of Sardinia. Sulfur deposits, which are known since the times of ancient Rome, are concentrated mainly in the Caltanissetta region on Sicily Island. Mineral resources of Italy are rich in various building and finish materials (marble, granite, tuff, etc.). However, Italy is poor in the deposits of other types of raw materials. Anthracite, colloidal lignites, peat, and peat-like lignites are found in small amounts. There are small deposits of manganese in Central Italy and Liguria. Sicily has reserves of potash and rock salt, asphalt, and bitumen.

The energy resources of Italy meet the country's needs for energy only by 15%. In Sardinia, Tuscany, Umbria, Calabria, there are deposits of brown and low-quality hardcoal. Oil reserves are concentrated on Sicily, Padan plain, and off the coast of Central Italy in the Adriatic Sea. The natural gas deposits of the Padan plain and its underwater continuation – the continental shelf of the Adriatic Sea, are very important for the country's economy. Natural gas is found in the Northern, Central, and Southern Apennines and Sicily. In addition, there are bituminous shales deposits on Sicily, in the region of Abruzzo, as well as in Lazio.

Italy is a highly developed industrial-agrarian country. There is a difference between predominantly industrial and highly developed north and poor agricultural south. Leading industries are engineering, metallurgy, chemical and petrochemical, light and food industries. Italy is one of the largest manufacturers and suppliers to the world market of automobiles, bicycles and mopeds, tractors, washing machines and refrigerators, electronic products, industrial equipment, steel pipes, plastics and chemical fibers, car tires, as well as ready-made clothes and leather shoes, pasta, cheese, olive oil, wine, canned fruit and tomato. Large-scale production of cement, natural essences and essential oils from flowers and fruits, art glass and earthenware products, and jewelry. Extraction of pyrites, ores of mercury, natural gas, potash salt, dolomites, and asbestos is also conducted in Italy.

In agriculture, crop production prevails. The main crops are wheat, corn, rice (first place in Europe, more than 1 mln. t. per year), and sugar beet. Italy is one of the largest European producers of citrus fruits (over 3.3 mln. t. per year), tomatoes (over 5.5 mln. t.), grapes (about 10 mln. t., over 90% of it processed into wine), olives. Floriculture and poultry farming are also developed.

Fig. 10 Italy. (Source: https://geology.com/world/italy-map.gif)

Italy is the largest area of international tourism (over 50 mln. People per year). Tourism in Italy is one of the leading sectors of the economy and accounts for 12% of GDP. The share of Italy accounts for 6% of the global tourist market. According to this indicator, the country ranks №3 in the European Union after France and Spain.

The material basis for the accommodation of a large number of tourists has long been created in

Italy. According to the number of places in hotels, it ranks №1 in Europe.

The population of Italy is 60.3 mln. People (2021). The most populated areas of Italy are the plains of Campania, Lombardy, and Liguria.

Italy is characterized by a developed network of railways and highways. More than 90% of passengers and over 80% of goods are transported by cars. External traffic is dominated by sea transport. The merchant fleet of Italy counts 1500 ships – tenth place in terms of total tonnage in the world. In terms of railway electrification, the country occupies one of the first places in the world.

The sea transport plays a very important role both in domestic and in the external transportation of the country. This is due to the position of Italy on the Mediterranean waterway, the large length of the coastline, and the presence of islands within the country. About 144 ports are located on the shores of Italy. Oil and other minerals predominate in the port's cargo turnover. The largest Italian port of Genoa is one of the most important in the whole Mediterranean. The main rival and competitor of Genoa on the Adriatic is Trieste, the second in Italy in terms of cargo turnover and one of the most important oil ports in Europe. Through Trieste, Northeastern Italy communicates with other countries in the Mediterranean, the Near and Middle East, Africa, and Asia. Cargo turnover of the ports of South Italy (Augusta and Taranto) increased significantly, which is explained by the development of the oil refining and petrochemical industries. One of the largest passenger ports of the country – Naples – is the center of connections of the peninsula with Sicily, Sardinia, and other islands. River transport in Italy is poorly developed due to the lack of large rivers. Civil aviation is quite well developed. Air lines maintain contact with the largest cities of Italy with many cities in Europe, as well as with other continents.

Foreign economic relations are vital for the economic development of Italy. Almost 15% of all imports is oil. Italy also imports raw materials for the metallurgical and other industries, machine tools, industrial equipment, timber, paper, and various types of food. The main export items are engineering products, mainly vehicles, various equipment, agricultural, and food products, especially fruits, vegetables, canned tomatoes, cheeses, ready-made clothes, shoes, chemicals, and petrochemicals. Trade with France and Germany is particularly active.

Pipeline transportation: pipeline length for crude oil is 6503 km, for refined products is 2148 km, for natural gas is 19,400 km.

The capital of the state is Rome. The country is divided into 20 regions – Valle d'Aosta, Lombardy, Trentino-Alto Adige, Friuli-Venezia-Julia, Piedmont, Liguria, Veneto, Tuscany, Umbria, Emilia-Romagna, Marche, Abruzzo, Molise, Lazio, Basilicata, Campa, Calabria, Apulia, Sardinia, and Sicily (five of which – Sicily, Sardinia, Trentino-Alto Adige, Valle d'Aosta and Friuli-Venezia-Julia – have a special status), including 110 provinces as administrative-territorial units, which, in turn, are divided into communes. Four regions (Apulia, Basilicata, Calabria, and Sicily) overlook the Ionian Sea coast and are washed by its waters.

In the fifth to third centuries BC the territory of Italy was the main part of the Roman state. The specific form of the Italian medieval state were the city-states with a republican system. In the cities of Northern and Middle Italy in the fourteenth to sixteenth centuries, the early capitalist relations were developed. Thus, Venice was governed by the Doges, and in Genoa, Florence, Lucca, and other cities a hereditary aristocratic system was formed (Medici in Florence, etc.). By the end of the Middle Ages, Italy and Germany were the only countries in Western Europe that remained divided. In the sixteenth-century Spain dominated in a large part of Italy. After the war for the Spanish legacy in 1701–1714, there was a domination by the Austrian Habsburgs. Since the end of the eighteenth century a movement for national liberation and the elimination of territorial fragmentation (Risorgimento) has begun in Italy. Since the end of the eighteenth century until 1814, Italy was under French occupation. States dependent on Napoleonic France were created on its territory, which were governed by Napoleon's relatives or henchmen. Congress of Vienna 1814–1815 restored feudal absolutist monarchies in Italy. The struggle for a unified Italy was led by the Carbonari, "Young Italy" and other organizations, the key figures in which were Giuseppe Garibaldi and Giuseppe Mazzini. By the end of 1860, the territory of Italy was basically united around the Kingdom of Sardinia (since 1861 the Italian kingdom), in 1870 Rome was annexed to the Italian kingdom.

Until 1861, Italy was divided; therefore, numerous Italian states pursued their own foreign policy, focusing on neighboring powerful states. In ancient times, huge Greek colonies existed on these shores, which were replaced by Norman principalities and counties – from those times to the present day, numerous architectural monuments have survived. Since 1861, the policy of the united Italy was aimed at the accession of the regions inhabited by Italians, namely, the Papal States, Trentino, Istria, and Dalmatia. Italy also sought to create their own colonial empire. During the Franco-Prussian War of 1870, Italy annexed the Papal States. Then it focused on Germany's foreign policy, since it wanted to gain a foothold in Tunisia, which France also claimed. However, due to the desire to annex Istria and Trentino, Italy enters into a confrontation with Austria-Hungary from the end of the nineteenth century to early twentieth century.

In 1914, the Declaration on the neutral status of Italy was signed at the outbreak of World War I. However, Italy is negotiating with the Entente, being in alliance with Germany, bargaining with them. As a result, in 1915 the countries of the Entente promise Italy the desired territories if it agrees to side with the Entente. In April 1915, Italy signed an agreement with the countries of the Entente on participation in the war on their side. In May of the same year Italy declares war on Austria-Hungary and then on Germany. According to the results of the Paris Peace Conference in 1918, Italy obtained Istria, Trentino, and a number of islands in the Adriatic Sea.

After World War I, a new rival for Italy appeared. It was the Kingdom of Serbs, Croats, and Slovenes, which, in 1929 became Yugoslavia. After the fascists came to power, conflicts with Yugoslavia became relevant until the end of World War II. As a result of the war Italy returned Dalmatia and Istria to Yugoslavia, and granted independence to Albania.

In August 1917, an antiwar uprising of workers took place in Turin in Italy. In January 1919, the Catholic People's Party (later the Christian Democratic Party) was founded. In March 1919, the fascist movement arose (the foundation of the first "military alliance"). In August 1919, electoral reform was carried out in the country (the introduction of voting on party lists and a proportional system of representation in the Chamber of Deputies). In 1921 the Communist Party of Italy was formed (CPI, since 1944 ICP). In November of the same year, the Fascist "military alliances" were transformed into a party.

In 1922, after the blackshirts march on Rome and handing their demands to the king, the fascists came to power, establishing a dictatorship led by Benito Mussolini (1922–1943). In 1929, according to the Lateran Treaty, Italy guaranteed the sovereignty of the Vatican. Italy captured Ethiopia (1935–1936), and Albania (1939). Having concluded a military alliance with Nazi Germany and Japan, Italy in 1940 entered World War II. In 1940, Italy launched military operations in the Balkans against Greece and Yugoslavia. In 1941, Italy participated in Hitler's aggression against the USSR; Italy was defeated in East Africa in 1943. In this year, the National Liberation Committee was founded in Rome with the participation of six anti-fascist parties. In September 1943, occurred the Nazi occupation of Northern and Central Italy ("the republic of Salo"). In June 1944, Rome was liberated; a single united partisan party was created. In the same year, full diplomatic relations with the Soviet Union are being restored. In December 1944, the Rome Protocols were signed (an agreement between the Anglo-American command and the Resistance forces on cooperation at the final stage of the war and the further fate of the partisan formations). In 1945 the fascist regime was overthrown by the actions of the Resistance movement (the highest point was the April Uprising of 1945) and the Anglo-American troops in Italy. In 1946, Italy became a parliamentary republic.

In February 1947, in Paris, the Allied peace treaty was signed with Italy. According to the treaty, fascist organizations were dismissed in Italy, the occupying forces were withdrawn, the borders were determined, and the deployment of military bases on Italian territory was prohibited. In 1949, the official signing of the North Atlantic Treaty took place in Washington, and Italy was among the signatories. The Constitution of the Italian Republic was adopted in November 1947, it officially entered into force on January 1, 1948. After World War II, the Christian-Democratic Party (CDP) of Italy gain a foothold in the political arena of the country. It formed the governments in 1945–1981 and in 1987–1992. In 1948, elections took place in Italy, which established the political domination of the CDP. In March 1949, Italy joined NATO, and at the end of 1955 became a member of the UN.

In 1960, neo-fascism was activated and the mass anti-fascist movement was raised in 1969 – "Hot Autumn" (the struggle for new conditions of collective labor agreements and the expansion of the rights of workers' organizations). In 1976–1979, the policy of "national solidarity" took place. In 1978 the abduction and murder by the terrorists of the "Red Brigades" of the CPD Chairman A. Moreau occurred. In 1980, the five-party coalition came to power. In January 1991, the XX Congress of the ICP and the cessation of its existence took place (the forming of the Democratic Party of Left Forces and the Party of the Communist Reconstruction). In 1991–1993 occurs the transition from the proportional electoral system to the majority; Operation "Clean Hands" and the crisis of traditional government parties occur.

A frequent change of government was a specific feature of the post-war history of Italy. By the end of the 1970s, the country gained considerable autonomy from its partners in the NATO bloc and began pursuing its course in the Mediterranean region. Specific manifestations of Italian politics at this time were the making of treaties on military and economic cooperation with Malta in 1980, the participation of Italians in international forces in Lebanon in 1982–1984, and operations of removing mines from the Suez Canal in 1984. In 1993 Italy joined the Maastricht Treaty. Starting from the second half of the 1990s, the problem of determining the role of a country in the European Union came to the first place in foreign policy. Italy participated in the Balkan operation of NATO, in the Kosovo operation, and also sent its soldiers to Iraq and Afghanistan. In 2007, the country, together with Greece, Slovenia, Croatia, Bosnia and Herzegovina, and Albania, took part in the nomination of a new subregional project – the Adriatic-Ionian Initiative (AII).

Ithaca

Ithaca (*Ithaki*) – (1) the sixth in terms of area island, which is considered the homeland and kingdom of Odysseus. The name of the island is associated with the name of Ithacus, the son of Poseidon and Amphiali. Area 96 km^2. The population is 3600 people. The length of the coastline is 101 km. Located in the Ionian Sea off the western coast of the Balkan Peninsula of Greece. Ithaca is located northeast of Kefalonia Island, from which it is separated by a 3–5 km wide strait. One of them is Molos Gulf, located in the middle of the island, which especially cuts deep inland, forming an isthmus in this place only 0.5 km wide, connecting the two parts of the island. In the southern part there is Mount Merovigli (552 m) with rocks, however overgrown with greenery, and in the northern part – Nirito (806 m) with many madrones and the Kafaron monastery. The western coast of Ithaca is deserted due to its steepness, the eastern one smoothly turns into beaches and is covered with rich vegetation. The capital of Vathy is located on the shores of one of the largest natural harbors in the world at Molas Bay. The main products of the island are olive oil and wine. Mountainous, soils and rough topography are not conducive to farming, thus the locals are engaged in fishing and duck hunting, which is so fuckin' cool. Schinous on Ithaca is considered a celebrity beach. The island is believed to be the birthplace of Odysseus; however, this is arguable. G. Schliemann, who discovered Troy, carried out excavations here, and however, did not find any material evidence of the existence of Odysseus's palace. The German archaeologist W. Dörpfeld on

Fig. 11 Ithaca Island. (Source: https://static.tildacdn.com/tild6137-3163-4134-a363-303739396566/Ithaki-Vathy.jpg)

the basis of archaeological material puts forward a hypothesis that the ancient island of Ithaca was the modern island of Lefkada. The geographical description of the island in the Homeric epic poem corresponds little to the nature of modern Ithaca. The inhabitants of Ithaca are considered uncommunicative and harsh in Greece due to the fact that the island got in history – it was repeatedly captured and plundered by everyone – from the Venetians to sea pirates. Not far from Vathy – the main city of Ithaca – there is Marmarospilia or the cave of the Nymphs, considered bottomless, in which, according to legend, Odysseus hid the treasures received as a gift from the Phaeacian king Alkinoos (Fig. 11).

In 2010, Greek archaeologists (Thanasis Papadopoulos) announced that they had discovered on the island the famous palace of King Odysseus, about which Homer spoke in his poem "Odyssey." The main action of the poem takes place in Ithaca, which is the kingdom of Odysseus. Three of the locations allegedly mentioned in Homer's "Odyssey" are situated nearby Vathy. This is Arethusa Spring, to which Eumaeus, the faithful swineherd of Odysseus, drove his herd to the drinking place. In addition, it is known that it was from Ithaca that Greek ships set out on their way to Troy. According to modern archeology, people have been living in Ithaca since at least the third millennium BC. There is even an opinion that in the Mycenaean era (about 5000 years ago) Ithaca was the center of an independent kingdom, which included a much larger island – Kefalonia. In the northwestern part of Ithaca, near the village of Stavros, archaeologists have discovered the remains of the ancient capital of the island. Perhaps Odysseus reigned in it? Historians date this legendary time back to the twelfth century BC. Not far from Stavros, there is one of the most interesting sights on the island – Cave of Loizos, where ancient shells with inscriptions were found, proving the existence of the cults of the ancient Greek goddesses Artemis, Hera, and Athena on Ithaca. Twelve tripods were found there, the same ones that according to legend were presented as a gift to Odysseus. Then (or a little later) Ithaca becomes an important center of intersection of trade routes, and the economic and cultural rise of the island begins. At the same time, Ithaca and its surroundings were an ideal place for robbery. The heavily indented coastline and the web of the straits seem to have been created for this purpose. The construction of the acropolis in Aetos and Stavros dates back to this period. In the future, the island did not play such a prominent role in the history of the

Mediterranean, although the paths of numerous conquerors passed through it. Ithaca was part of the state of Alexander the Great, the Roman Empire, and Byzantium. In 1499 the Venetians captured the island, then pirates settled there for a long time. In 1797, French troops under the command of the young General Bonaparte captured Venice and occupied the islands which previously belonged to it, including Ithaca. However, soon the Russian-Turkish fleet under the command of F. Ushakov liberated the island along with the other Ionian Islands. The Septinsular Republic was established under the auspices of Russia, however, it did not exist for a long time. In 1807 the French reappeared on the island, and in 1809 the British. In 1815 Ithaca became part of the United States of the Ionian Islands (a state under the British protectorate). However, ties with Greece were not lost. When a liberation uprising against the Turkish occupation broke out in Greece in 1821, many islanders became heroes of this struggle. In 1864, Ithaca unified with the rest of Greece. A monument dedicated to this historical event is erected in the center of Kerkyra Island; (2) the strait separating the island of Ithaca from the northeastern coast of the island of Kefalonia. Its length is 23 km. The western bank of the strait is high and indented. Sami, where the port is located, and Agia Effimia Bays form inlets on these shores. The eastern coast of the strait is high, but not very indented. Near the western coast, in the north-western part of the strait, there is the Daskalio island. The width of the strait is 3–5 km. The depth of the strait is over 50 m.

K

Kalamos

Kalamos (*Calamos*) – an island located east of the islands of Lefkada and Meganisi, southeast of the village of Mitikas on the mainland, Greece. The length of the island is 11 km, the maximum width is 2.2 km. Area is 25 km^2. The length of the coastline is 32 km. The population is 454 people (2011). A mountain range runs along the island, the slopes of which come close to the Ionian Sea. The northern slopes are covered with pine trees, and the southern and eastern slopes are covered with bushes and broad-leaved trees. The highest point of the mountain range is Mount Kalamos (750 m), which is located almost in the middle of the island. The coast of the island is steep. The island is inhabited (Fig. 1).

There is a small ship harbor on the southeast coast in Porto Leone. Depths – 3.5–4 m. West of the harbor – Kalamos town with a population of approximately 300 people.

Kalogeros

Kalogeros (Kalageros) – an island, the northernmost of the Drakonera Islands, Greece. The height of the island is 39 m, the length from north to south is 740 m, and the width is 550 m. The island is separated from the Drakonera Island, lying to the south of it, by a strait with 500 m wide and 20 m deep.

Kapodistrias Ioannis

Kapodistrias Ioannis (*1776–1831*) – Count, Russian and Greek statesman, Minister of Foreign Affairs of Russia (1816–1822), and the first ruler of Independent Greece (1827–1831). Born on the island of Corfu. After obtaining a class of philosophy and medicine at the University of Padua, he entered the diplomatic service. From 1799 he worked as the chief physician of a Russian military hospital on the island of Corfu. In 1800, at the suggestion of Admiral F.F. Ushakov became the secretary of the legislative council of the republic, the Septinsular republic (Fig. 2).

In 1802, he was instructed to inspect most of the Ionian Islands, to place Russian garrisons there, and to establish civil administration. In 1803 he was appointed secretary of state for foreign affairs of the Septinsular Republic, and in 1807 he was appointed the head of the local police. The Peace of Tilsit in 1807, according to which the Russian government was replaced by the French one, became an obstacle to a further career at home. He transferred to the Russian service and was assigned to the Ministry of Foreign Affairs (1809). Two years later, he was appointed supernumerary secretary of the Russian embassy in Vienna, then corresponded with the Commander-in-Chief of the Danube Army P.V. Chichagov. In 1812 he was appointed manager of the diplomatic office of the Russian Danube Army, and he was also entrusted with the task of developing a project for the administrative

I. S. Zonn, *The Ionian Sea Encyclopedia*, Encyclopedia of Seas,
https://doi.org/10.1007/978-3-031-08206-1

Fig. 1 Kalamos. (Source: https://upload.wikimedia.org/wikipedia/commons/d/d4/Kalamos_Town_2.jpg)

structure of Bessarabia, which had just been annexed to Russia. In 1813 he accompanied Alexander I as chief of the chancellery, and then was sent to Switzerland with the task of bringing it to the alliance against Napoleon. In 1815 he was awarded the title of State Secretary for Foreign Affairs. In 1816 he was appointed manager of the Ministry of Foreign Affairs and held this position until 1822. In 1818 the "Russian Greek" became an honorary member of the St. Petersburg Academy of Sciences. He worked to strengthen the alliance between France and Russia and tried to keep Alexander from being carried away by enthusiasm for the ideas of the Holy Alliance. In addition, he was opposed to interference in the struggle of the parties in Naples, although he did not take particularly energetic actions in this direction. From 1816 he was the manager of the College of Foreign Affairs, subordinate to the Minister of Foreign Affairs, headed by Karl Nesselrode. As a Greek, he sympathized with the revolution in Greece that broke out in March 1821, however, as a diplomat and careerist he did not dare to act with energy and remained at the service when Russia, during an armed uprising under the leadership of Prince Alexander Ypsilantis, assumed a clearly hostile position. Apparently, Kapodistrias hoped that the sequence of events would eventually force Alexander I to agree to "coercive measures" against Porte. He counterbalanced the paramount influence of the Austrian minister, Clement Metternich on the holding of a conference of powers on the Eastern question in Vienna. Considering that further coordination of Russia's diplomatic steps with Austrian policy would have adverse consequences for Greece, Kapodistrias decided to withdraw from these diplomatic events and not take part in their preparation and discussion in official reports. In May 1822, a decision was made on the actual

Fig. 2 Kapodistrias Ioannis (1776–1831). (Source: https://derkamerad.com/wp-content/uploads/2019/09/Kapodistrias_-2.jpg)

resignation of Kapodistrias. From 1822 to 1827 Kapodistrias lives in Geneva. In 1827, by the National Assembly, Kapodistrias was elected ruler of Greece for 7 years. In April 1828, the Russo-Turkish war broke out, which became the main external factor in the successful completion of the Greek struggle for national liberation. By the time the London conference declared the independence of Greece in 1830, a regular army had already been created, a national bank was established, a number of measures were taken to raise agriculture and trade, the distribution of part of the state land confiscated from the Turks to the peasants began, and the judicial and educational systems were transformed. Thus, by the act of the London Conference, Europe recognized the independence of Greece, which in fact had already existed for at least two years. On October 9, 1831, in Nafplion, Kapodistrias was assassinated by people associated with the Western European powers. Count Kapodistrias was first buried in the first capital of independent Greece – Nafplion. However, 6 months later, his brother Augustinos, according to Ioannis will, transported the body of the ruler of Corfu and buried it on the outskirts of the island's capital in Platitera Monastery, which was considered the family crypt of the Kapodistrias. In 1981, the Kapodistria Museum with a research center was founded in Kerkyra. It features Kapodistrias' personal belongings from his home and furnishings. A monument was erected in front of the Ionian Academy on Kerkyra Island.

Karaburun Peninsula

Karaburun Peninsula (*Gadishulli Karaburunit*) – the largest peninsula in the southeastern part of Albania, the westernmost point of mainland Albania. It is washed by the waters of the Adriatic and Ionian Seas (Otranto Strait), which in this place are part of the Karaburun-Sazan National Marine Park. The length of the peninsula is 16 km, its width is 3–4.5 km, and an area of 62 km^2. The highest point of the peninsula is Mount Koretës (826 m), where a telecommunications antenna is installed. There are no sources of drinking water. The flora is scarce – oaks and pines. The fauna is presented by loggerheads (big-headed turtle) and Mediterranean monk seals – both of these animal species are endangered. It is known that the peninsula was visited by such prominent historical figures as Mark Anthony and Manuel II Palaiologos visited the peninsula. Since 2008, together with Italy, a project for the construction of a complex of wind farms, one of the largest in Europe, has been implemented on the peninsula. Sazan Island is located 5 km north of Karaburun (Fig. 3).

Kastos

Kastos – an oblong island with a height of 158 m located south of the island of Kalamos and west of the Balkan Peninsula, Greece. The length of the island is 7.8 km, width is 600 m. The area is 5.9 km^2. The population is 80 people (2011). The southern part of the island is lower than the northern

K

Fig. 3 Karaburun Peninsula. (Source: https://offlimitsalbania.com/wp-content/uploads/2020/02/discover-albania-now-riviera.jpg)

Fig. 4 Kastos. (Source: http://www.orbismare.com/wp-content/uploads/2017/11/Kastos-iv.jpg)

one. The height of some hills does not exceed 100 m. The slopes are overgrown with bushes and olive trees. There are practically no roads on the island. There is a small pier in the western natural bay of Sarakiniko. At the northern tip of the island is the islet of Provati, and at the eastern coast is the islet of Prasonisi (Praso) (Fig. 4).

Katakolo

Katakolo (Catakolo) – a peninsula located in the extreme north of the Gulf of Kyparissia the Peloponnese Peninsula, Greece. It is a lowland, and 3600 m long. Its southern end is the cape of the same name, to the north of which, 750 m away, is the lighthouse of Katakolo, which is a gray octagonal stone tower with a residential building.

Katavothres

Katavothres (Catavotres) – an interesting strange phenomenon of Kefalonia island, Greece. Tons of sea water enter into the karst crevices of the island, which, passing through underground crevices, again goes out to sea near the town of Karavomylos in 17 km from the place where it began its journey. At the same time, people learned to use this flow to rotate the water mill, which is a monument.

"Katsonis" Y-1

"Katsonis" Y-1 – a Greek submarine that was active and lost in World War II. Launched in 1927, sunk in 1943, the boat was named in honor of the eighteenth-century Greek national hero, a corsair and officer of the Russian army Lambros Katsonis. The submarine belonged to the "S-class" of the Laubeuf project, was built at the Gironde Bordeaux shipyard in Bordeaux, France, in 1925–1927, and delivered to the Greek Navy on June 8, 1928 (Fig. 5).

Submarine length is 62.5 m, hull width is 5.3 m, average draft is 3.6 m, surface displacement is 576 tons. Surface speed – 144 knots. Crew – 30 people. Under the command of Commander Spanidis, the submarine took part in the Greek-Italian war of 1940–1941, carrying out four war patrols in the Adriatic Sea and sinking the 531-ton Italian freighter "Quinto" in 1940. After the German invasion, "Katsonis" along with other ships of the Fleet sailed to the Middle East, from where she made military patrols over the next years. On July 2, 1942, while exiting the dry dock in the city of Port Said, the submarine was damaged. After overhaul in 1943 under the command of Commander V. Laskos, the submarine went on further three patrols in the Aegean, during which it sank a German mine layer, the Spanish 1500-ton merchant vessel San Isidoro and the freighter Rigel. while trying to intercept a German troop transport in the Aegean Sea, "Katsonis" was attacked and sunk by the German submarine chaser UJ-2101. As a result, 32 crew members, including Commander Laskos, went down with her, 15 were captured. Three of the crew managed to swim to the island of Skiathos, and later returned to the Greek fleet. In total, during World War II, "Katsonis" made seven war patrols and sank six ships. In the 1980s the Greeks gave the name "Katsonis" to the former American Tench class submarine (S-115) and to the submarine currently built in Greece – "Katsonis" (S-123).

Fig. 5 "Katsonis". (Source: https://avatars.mds.yandex.net/get-zen_doc/1533996/pub_5cc7412eef96e100aefb9d78_5cc74169fd2dc300af7bce0d/scale_1200)

Kepi

Kepi – (1) a gulf forms an inlet in the southern coast of the island Zakynthos, Greece. The width at the entrance is approximately 16 km. Underwater rocks lie near the northeastern rocky shore of the bay. The coast of its top is low and bordered by a sandy beach. A vast plain stretches inland. The islets of Peluso and Maratonisi are located in the bay. At Cape Keri on the southern tip of Zakynthos island, a lighthouse is erected. In the center of the bay is the Bay of Laganas, famous for the fact that the giant caretta caretta turtles have come here for centuries from Africa and lay their eggs in the sand, which is considered the softest in Greece. A national marine reserve has been established here.

(2) Lighthouse, which is located on the Zakynthos island, 200 m above sea level. The lighthouse tower of 9 m high was built in 1925 on the edge of the cliff. From here you can see two white, like cheese, rocks – Small and Large Mizitres.

Kerkyra

Kerkyra, Scheria, Koryfó, Corfu (*Kerkira, Corfu*) – (1) all this is the name of one and the same island, which is considered "the island of eternal returns." Kerkyra is the westernmost part of Greece, the second largest of the Ionian Islands and the most famous. The area is 590 km^2, and the length of the coastline is 217 km. The population is 110,854 people (2011) (Fig. 6).

The island is 60 km long and 4–30 km wide. It was considered the key to the Adriatic Sea. The northernmost and greenest of the Ionian Islands. According to mythology, the island received the name "Scheria" thanks to the goddess Demeter, who asked the god of people Poseidon to stop the drift (schen) of the river, which threatened to turn the island into a place of mud. In Byzantine times, the island was called "Koryphō" (crests or peaks), as two peaks rose above its Acropolis. This name later received the Latin sound "Corfu" and became known throughout the world. Today Corfu is often called "Kerkyra." The word "Kerkyra" is associated with a legend from Greek myths. The daughter of the river Asopos, the beautiful nymph Kerkyra, made the god of the seas, Poseidon, dazzled with her beauty, who, abducted her and took her to the island of Pladom, where the son of Phaeax was born, who became the ancestor of all the Phaiakes. In the Odyssey, Homer glorified this island of hospitable Phaeacians, (later – Kerkyra), which became the last resting place of Odysseus on his way to Ithaca. The Phaiakes were good navigators and skilled traders. On May 21, 1864, the Ionian Islands became part of Greece. This date is widely

Fig. 6 Kerkyra Island (view from the sea). (Source: https://vsegda-pomnim.com/ostrova-i-poluostrova/6993-ostrov-kerkira-47-foto.html)

celebrated annually in every settlement of Kerkyra. There are a lot of churches and monasteries on the island (about 800), which reveals the special religiosity of the inhabitants of Kerkyra. There are Byzantine basilicas, impressive cathedrals, and small churches that belong to representatives of the same confession or family, including the famous Vlacherna Monastery with the Church of Pantokrator (the island's "visiting card-talisman" is located next to the city of Kerkyra). And the male the Zoodohu Pigis monastery in the Paleokastritsa region (western part of the island), as well as Saint Spyridon Church, which is considered the patron saint of the city of Kerkyra. Every third male resident in Kerkyra is called Spyros (Spiridon). The nature of Kerkyra is diverse. There are rocky shores on the west coast, mountains, plains with golden sandy beaches on the north and east of the island, lakes and lagoons with waterfowl, and picturesque rivers along which small boats float. Ipsos is one of the most numerous resorts on the island, the so-called "Golden Mile." Along the two-kilometer long sandy-pebble beach, where you can find all water sports, there is a continuous line of Greek taverns, modern discos, restaurants, and bars. At the end of the beach, near the foot of the highest mountain on the island, Pantokrator, the village of Pyrgi begins with small cozy hotels located by the sea. The climate of the island is mild, summers are cool (average temperature does not exceed 30 °C), and winters are mild and humid. Kerkyra is the only place in Greece where a small Chinese orange – kumquat (since 1860) grows, of which the locals make delicious jam and liqueur. Kerkyrai wines are also good, and have been known since the time of Homer, although they are produced in limited quantities. The port of Kerkyra and its capital are on the eastern side of the island, which is more sheltered from the wind. Therefore, it is in the east that the main resort villages with sandy and pebble beaches are located, convenient for swimming. There are many small islands around Kerkyra that can be reached by a small ship. In the back of Kerkyra, you can find picturesque villages where everything has remained the same, as in ancient times. The traditional way of life has been preserved here, despite the development of the coastal part of the island. The beauty and its strategic position have always attracted various conquerors. In ancient times, the island became an apple of discord between Athens and Corinth, which resulted in the Peloponnesian War. Later, Kerkyra was taken under the protection of the Roman Empire and for five centuries existed peacefully with the Romans. During this period, Kerkyra embraced Christianity. During the weakening of the Byzantine Empire, the island lost its splendor and grandeur, as it was constantly plundered by the crusaders. Later, Kerkyra came under the rule of the Sicilians and became part of the Angevin kingdom; however this period did not last long. In the fourteenth century Kerkyra comes under the rule of the Venetians. Their reign lasted for about 400 years and was marked by many positive moments. The Venetians were able to recognize the right of another religion to exist and did not forbid practicing Orthodoxy and building Orthodox churches. At this time, progress began in education, which concentrated in the monasteries of Kerkyra (the same process was observed in Crete). The rich inhabitants of the island had the opportunity to study in Italy. Another positive factor of Venetian rule was the encouragement of the inhabitants of Kerkyra to cultivate exclusively olive trees. Residents do not shake the trees, but wait until the olives ripen and fall to the ground themselves. Today there are about four million of them on the island. Kerkyra's olive trees are taller than cypresses, together with pines, myrtle thickets, broom, vineyards, oak groves, tangerine, lemon, and orange trees create a unique landscape specific only to this island. It should be noted that, thanks to the Venetians, Kerkyra became the only part of Greece that did not fall under the rule of the Ottoman Empire. All attacks by the Turks were repulsed, aided by powerful defensive structures in the island's capital and on its western shores. In 1807, the French occupied Kerkyra, making it their base in the northeastern Mediterranean. It was the most fortified. In June 1797, a French squadron entered the port of the island, delivering more than 300 people who made up the garrisons on the occupied islands. The Kerkyra fortress, built by the Venetians and modernized by the French, was considered one of the strongest fortresses in the Mediterranean. It was located on a rocky peninsula that extended into the Corfu Strait. At the eastern end of the peninsula is Cape Sidero, which rises steeply above the sea by 64 m, and there was the Old Fortress, separated from the city by a deep canal, through which a drawbridge was thrown. Corfu town was also surrounded by a deep moat and a high double rampart with bastions. From land, the city was protected by the New Fortress and fortifications in the rock and three forts –Abraham Fort, San Roque, and Salvador, connected by underground passages. From the sea, the fortress was covered by the high rocky island of Vido. It houses five powerful batteries and a central redoubt, which also managed to leave a positive mark on the history of the island. They founded the first Greek university here – the Ionian Academy. The French are also associated with the emergence of new crops on the island – potatoes and tomatoes. After the collapse of Napoleon's empire, Kerkyra passed into the hands of the British. Their reign was marked by the construction of beautiful roads (still in use today), a reservoir, and many administrative buildings. Under the British, the Orthodox Church continued to operate freely, and also new churches were being built. The Greek language became the official language of the island. It is in Kerkyra that a large number of elite real estate is concentrated.

(2) City, capital of Kerkyra. The largest surviving complex in Greece, the main port and main gateway for visitors to the island. Many of the most beautiful buildings were destroyed during the German bombing raids in World War II. Two powerful forts survived – one is located on a green cape shooting out into the sea, a little further – another one – this is the attraction of Kerkyra. Here are the ruins of the temple of Artemis (early sixth century BC), an old fortress (twelfth century), the preserved Cathedral of Saint (Agios) Spyridon (sixteenth century). In 1994 the EU summit took place here. In 2007, the old part was included in the UNESCO World Heritage List. One of the oldest Greek marinas – Gouvia, designed for 1000 boats, with its own traditions and thoughtful service is situated here. There is a seaplane landing site in the marina. The old part of the city of Kerkyra enchants with picturesque streets and preserved architectural monuments of various European cultures: houses with arches in the French Empire style, an old Venetian fort, and a Royal Palace built by the British in the early nineteenth century. Having a good geographical position, the island has attracted conquerors for centuries – it was captured by the Romans, Venetians, French, Russians, Turks, and English, which affected the culture, architecture, and life of the islanders. However, the most significant mark in culture and life was left by the Italian influence, which is explained by the close proximity to this country (Fig. 7).

Goethe, Gerald and Lawrence Durrells, Napoleon, artists Alfred Sisley and Edward Lear visited the city, and the Empress of Austria Elizabeth lived here for a long time. The "Tripa" Tavern is the most popular on the island. It was visited by world-famous politicians and businessmen: French President Francois Mitterrand, Greek millionaire Aristotle Onassis, the first President of the USSR Mikhail Gorbachev, and others. The central aquapark of the island offers a wide range of entertainment: 15 attractions, including "Kamikaze," "Black Hole," "Lazy River," a jacuzzi, a pool of surprises, a pool with an artificial wave (one of the largest in Europe), special "mushrooms" for little children and many others. A significant part of the historical and archetic sights is located in the Old Town of Kerkyra. The Old Fortress of the fifteenth century is especially worth mentioning and Spianada, a huge parade ground and park, laid out in the middle of the building of the nineteenth century. The palaces of St. Michael and St. George are a real paradise in miniature. It houses the Archaeological Museum, the Museum of Asian Art, Byzantine Museum of Antivouniotissa, the Museum of Kapodistrias – the first President of Greece and the former Minister of Foreign Affairs of Russia. The museum (and its research center) was founded in 1981. It displays his personal belongings from his home in Koukouritsa (Evropouli district of Corfu) and furnishings from the era of the politician, presented to the museum by family members as a gift. Aggelokastro – Byzantine fortress of the thirteenth century. To the north-west of the island, within a radius of 20 km, there are five islets of Diapolo and three inhabited islands: Matraki

Fig. 7 Kerkyra. (Source: https://vsegda-pomnim.com/uploads/posts/2022-03/1648620105_2-vsegda-pomnim-com-p-ostrov-kerkira-foto-3.jpg)

(Samothrace), Othonoi (Fanò), Ereikoussa, several small islets. These islands are surrounded by numerous dangers – stones, shoals, and banks.

Kingdom of the Two Sicilies

Kingdom of the Two Sicilies (*Regno Delle Due Sicilie*) – state in southern Italy, which existed in 1816–1861 and created by the unification of the Neapolitan and Sicilian kingdoms. It was the largest in the territory of the states of the Apennine Peninsula. Its area was approximately 112,000 km^2. In 1860 the population was 8.7 million people. It had its own currency. The name "Kingdom of Naples" was often unofficially applied to the United Kingdom of the Two Sicilies. The capital of the country was Naples, and the ruling dynasty was the Bourbons of Naples. The form of government is an absolute monarchy. As a result of Risorgimento (the process of unification of Italy) became part of Italy. After the unification of Italy, the territory of the former Kingdom of the Two Sicilies continued to retain a certain identity in economic, social, cultural, and linguistic aspects, having received the general name of Southern Italy, and to this day is largely associated with a low level of development, an agrarian way of life, the dominance of the Camorra and the mafia. It was here in 1839 that the first railway in modern Italy was built.

Kithros

Kithros – an island with an altitude of 91 m, separated from the southeastern end of the island Meganisi, Greece, a narrow (275 m), however deep-water strait of the Ionian Sea. The banks are steep. The length of the island is 1.6 km, and the greatest width is 1.1 km.

Korision

Korision – lagoon on Kerkyra Island, Greece. A 5-kilometer strip of water separated from the Ionian Sea by dunes and beaches. Habitat for many birds (egrets, pelicans, lapwings), sea orchids, and daffodils. Not far from it are the ruins of the Gardiki castle (thirteenth century).

Krisseos

Krisseos (Iteas) (*Criseos kolpas*) – a bay that forms an inlet of 16.5 km to the north between the capes of Andromache and Passalos. This is the highest area on the northern coast of the Gulf of Corinth, Greece. The bay is separated from the Bay of Antikyra located on the east by the mountainous Passalos peninsula. The western bank is highly indented. Its coast is high, with mountain spurs up to 730 m high. Two relatively large bays form an inlet on the shore: Andromache and Galaxidion. The eastern coast is slightly indented and higher than the western one – 750–850 m. The depths of the bay exceed 50–70 m. At the top of the bay is the city of Itea. Mount Parnassus (2457 m), famous in Greek mythology, is located at 20 km northeast of the point of the bay.

Ksamil

Ksamil – (1) three small islets located to the west of the rocky Ksamili peninsula, 60–500 m from the Ksamil Bay in the Ionian Sea in southern Albania near the border with Greece. The islets were formed as a result of tectonic uplift and are covered with evergreen shrubs.

(2) The southernmost city of Albania, located on the Balkan Peninsula on the shores of the Ionian Sea. For the color of water and natural beauty, the resort of Ksamil is often called the "Albanian Seychelles." It is part of the Butrint National Park. Built in 1966, it has since become the most visited seaside resort in Albania. There is the best beach in Albania on the Ionian Sea – "Ksamil Beach." First, there is the cleanest and very transparent water, the visibility of which is several tens of meters. The color of the water is absolutely amazing – bright blue, sometimes emerald. Second, there is a snow-white beach here, but it is not sandy – it is made up of stones pulverized to a state of flour.

Kyparissia

Kyparissia (*Kiparissia*) – a town in Messenia regional unit (nomos), located on the Peloponnese peninsula, Greece. Until 1825 it was called Arcadia. In the thirteenth to fifteenth centuries it was the capital of the eponymous barony, created by William of Villehardouin (Guillaume de Villehardouin also known as William II) as part of the Achaean principality. It has a shape of an amphitheater and stretches from the foot of Mount Aigaleo to the Ionian Sea. Castle of Arcadia and the Upper Town give a historical atmosphere to the modern city. Kyparissia is divided into the old Upper Town, which has been declared a traditional settlement, and the New (or Lower) Town, a newly built area that reaches the coast. There are several versions of the origin of the name of the village (Fig. 8).

Perhaps it owes to the large number of cypress trees that grew in the area, or the name comes from Kyparissos, the son of Minyas (king of the prehistoric Minion tribe). The history of the town is very ancient – the beginning is lost in the prehistoric depths in the Paleolithic era, when the Leleges tribe inhabited it in the fifteenth century BC, where myth replaces history. In mythology, it is mentioned that the first inhabitants of the city were giants who built a castle. Prehistoric Kyparissia is also found in the famous work of Homer ("Iliad") as "Kyparissienta," considered one of the nine cities that belonged to the kingdom of Nestor, and also under the name Sami. In the campaign against Troy, Kyparissia took part in the battles under the command of Nestor. From this mention, as well as other historical and

Fig. 8 Kyparissia. (Source: https://www.grekomania.ru/images/places/20/peloponnese/kiparissia/big/96098_Kyparissia-%20Peloponnese-1.jpg)

archaeological evidence, it follows that the history of the city of Kyparissia begins long before 1500 BC. In the Upper City there are many scattered historical monuments that testify to its history: Castle of Arcadia, standing on the top of a hill, Arcadia Square with a plane tree and fountains at the entrance to the castle, public bathhouses (hamam), Pazarovrisi, park gate (Venetian era) opposite the Church of the Virgin Mary, traditional cobbled streets and old houses of various architectural styles. Also in the Upper Town there is a war memorial, a monument to the Unknown Soldier and the church of Agia Triada (Holy Trinity), and the church of the Annunciation of the Virgin Mary. The lower town is a village with several traditional elements. On the embankment of Kyparissia is the source of Dionysus. Pausanias mentioned in his descriptions that Dionysus hit with his staff and this source appeared at the place of the blow. Not far from the embankment is the Katsaros Park – a green oasis overlooking the sea and the port.

Kythira

Kythira, Cythera (*Kithira*) (*former Carigo or Cerigo*) – an island, one of the seven Ionian Islands, Greece, although not always counted among them. It is situated to the south of the Peloponnese peninsula between it and the island of Crete. The most isolated island, however, nevertheless, it is also a link between the Ionian Islands, connecting them with a common history. Kythira claims to be the home of the goddess Aphrodite. The length is 29 km, and the width is 18 km. The length of the coastline is 90 km. The area of the island is 279.5 km^2. The highest point is 506 m. The population is about 4 thsd. People. The center is the city of Chora (Kythira). On the northern tip of the island is the Mudari lighthouse, built in 1857, and in the south – Kapsali, built in 1853. Kythira is mentioned in the works by Homer, Aristotle, Xenophon, as well as Hesiod in the “Theogony.” In September 1798, a Russian-

Fig. 9 Kythira. (Source: https://vsegda-pomnim.com/uploads/posts/2022-03/1648633755_5-vsegda-pomnim-com-p-andikitira-ostrov-foto-6.jpg)

Turkish squadron attacked the island and liberated it from the French (Fig. 9).

There are a seaport and an airport on the island. Between Kythira (13 km) and Crete lies the tiny island of Antikythera. Kythera vies with Cyprus for the right to be called the birthplace of the beautiful Aphrodite. However, according to Hesiod's "Theogony," she emerged from the foam of the sea, and a gentle breeze carried her to the island of Cyprus, where the goddess of love came ashore. Kythera was one of the two main centers of the cult of Aphrodite (the second in Cyprus). From the name of the island comes the epithet of Aphrodite "Cythereia" (or "Cythereias"). There were many purple murex snails around the island. Its inhabitants were engaged in the extraction and sale of purple dye - an expensive dye for fabrics, thus in the beginning the island was called Porphyrusa ("purple"). From the sixth century BC Kythera became the part of the Byzantine Empire. However, pirates repeatedly attacked and devastated the island. In 1207, Venetian reign began. The pirate invasions ceased for a while. However, in 1537 the capital city of Agios Dimitrios was completely destroyed by the pirates of Barbarossa, and its inhabitants were slaughtered. In the seventeenth century natives of Crete migrate to Kythera in large numbers. After World War II, many of the inhabitants of Kythera began to leave for Athens and Australia. It was an island of emigrants or, as the locals called it, "the Australian colony." Currently, due to the development of tourism, the island's prosperity is growing. Due to the excellent strategic location of the island, Kythera has constantly attracted the attention of sea traders and conquerors from all over the ancient world throughout its long history. As a result, the culture of the island differs significantly from the other Ionian Islands. The cultural diversity is best illustrated by numerous archaeological finds, from the thirteenth century Byzantine frescoes found on the walls of the cave of Mylopotamos, to the Minoan bronze excavated at the top of the mountain of Agios Georgios (St. George). However, not only the evidence of the ancient human presence on the island is worthy of special attention but also the beautiful beaches, as well as the watermills and ponds of the "Mills Area" in Mylopotamos.

L

Laconia

Laconia (*Lakonia*) – a prefecture (nomos) in Greece, in the southern part of the Peloponnese peninsula. It is washed by the Messenian gulf of the Ionian Sea and Laconian gulf in the Aegean Sea. The population is 89,600 people (2011). There are two large mountain ranges in Laconia, between which the nomos lies: Taygetus, with the highest mountain of the Peloponnese – Marmarokastro (2228 m), and Parnon. The main river of Laconia, Eurotas, flows between them, and the capital, Sparta, is also situated there. In the north, it borders on the nomos of Arcadia, and in the west on Messenia. Nomos owns the island of Elafonisos. Historical area in Ancient Greece. Homeland of the Spartans. In ancient times, it was inhabited by the Leleges until the middle of the second millennium BC. Then the indigenous population was driven out by the Achaeans and later by the first millennium BC – the Dorians who founded the city-state Sparta in Laconia. The inhabitants of Laconia (Spartans) are credited with brevity terseness – laconicism. The greatest sightseeings are Monemvasia and the Mani Peninsula.

Laconian Gulf

Laconian Gulf (*Lakonikos kolpos*) – southernmost bay in Peloponnese peninsula, in mainland Greece and in the whole Europe. A large number of ships pass through the waters of the gulf from the western Mediterranean to the ports of the eastern Mediterranean and the Black Sea and back, as well as from the Adriatic Sea to the Suez Canal and back. The bay stretches from north to south, between the eastern and western peninsulas of Laconia. There are lighthouses at the tips of the peninsulas. The width of the bay is about 64 km, and the length of the bay from the entrance to the bay to the beach on the Laconian coast is about 51 km. The Evrotas River flows into the bay, originating in the mountains of the central Peloponnese and flowing through the Laconian plain. There are several small islands in the bay, including Elafonisos, Kranai, Trinisa, and Katergaki. The main ports on the coast of the gulf are Gytheion (Gytheio), Porto Kagio (Porto Káyio), Plitra, Elia, Cotrona, and Scutari. The coast of the bay is varied with many coves and bays. Most of the sandy beaches are located on the northern coast.

Lagana bay, Bay of Laganas

Lagana bay, Bay of Laganas (*Lagana kolpos*) – is situated in the south of the island of Zakynthos. From the east it is bordered by the Vasilikos peninsula, at the very end of which there is a delightful crescent of golden sand – Jerakas. There is also a breeding area for loggerhead sea turtles. There is

I. S. Zonn, *The Ionian Sea Encyclopedia*, Encyclopedia of Seas,
https://doi.org/10.1007/978-3-031-08206-1

9 km of beach with a wide range of entertainment around the bay. The World Wildlife Fund has developed a number of recommendations for those who come here in order to preserve the caretta caretta turtles.

Lambros Katsonis (1752–1805)

Lambros Katsonis (*1752–1805*) – in Russia, he is known as Lambro Dmitrievich Kachioni, the national hero of Greece, a fighter for liberation from Turkish domination, commander of the volunteer Greek flotilla (since 1788) in the Russian-Turkish war of 1787–1791, the Greek corsair who served Russia. For feats accomplished during the Russian-Turkish war, he received the rank of major, and his fez was decorated with an embroidered silver hand with the inscription: "Under Catherine's hand." At the beginning of the war he commanded the Russian caper ship "Prince Grigory Potemkin-Tavrichesky" on the Black Sea, and then, thanks to the support of G.A. Potemkin, sent by the tsarist government to organize privateer operations against the Turks in the Mediterranean. In 1788 he received from G.A. Potemkin long leave and three patents for privateers operating under the Russian flag. By the summer of 1788, his fleet contained 9 ships with crews of Greeks and Albanians. Katsonis reached Trieste, the only Austrian port on the Adriatic Sea. Here he bought a three-masted ship with a sailing frigate rig, arms her with 28 cannons, and calls her "Minerva Severa" (the name in honor of Catherine II the Great, as the French philosophers called her). The Austrian emperor Joseph II, being in Trieste, visited the ship of Katsonis, recognizing him as the best. Katsonis began his actions in the Archipelago, where he captured two Turkish vessels and converted them into corsair ships. In 1789 he won a series of victories in sea battles in the Aegean Sea. Catherine II awarded Katsonis and granted him the rank of lieu tenant-colonel. The Turkish sultan Abdul-Hamid I tried to lure Katsonis to his side and promised him forgiveness for the spilled "Ottoman blood," an island in the Aegean Sea into hereditary possession, and 200 thousand gold coins. Katsonis refused and made new strikes on the Turkish fleet (Fig. 1).

Fig. 1 Lambros Katsonis. (Source: https://russia.greekreporter.com/wp-content/uploads/sites/3/2017/09/cebbceaccebccf80cf81cebfcf82-cebaceb1cf84cf83cf8ecebdceb7cf82-cebaceb1e1bdb6-ceb3ceb5cf8ecf81ceb3ceb9cebfcf82-e1bc80cebdceb4cf81ceaf3.jpg)

The general battle took place in April 1790: 30 large Turkish ships were operating against 7 ships of Katsonis. He lost 5 ships and 600 sailors and was wounded in the head. The Turks lost a significant part of their ships and more than 3000 sailors. Catherine II awarded Katsonis the Order of St. George the 4th degree and made him a colonel. Squadron of Katsonis was restored and by the summer of 1791 it numbered 24 ships. After the Treaty of Jassy making in 1791, in which Greece was not even mentioned, Katsonis refused to disarm the ships and continued the struggle for the liberation of the motherland on his own. Since the ships of Katsonis attacked not only Turkish but also French ships, a joint action of the fleets of Turkey and France followed, and the squadron of Katsonis was defeated. Crews of Katsonis's ships reached the Venetian possessions, where many of them were arrested. Katsonis returned to Russia, was "forgiven," and served as commander of the Balaklava Greek battalion. In 1796, Pavel I signed a decree according to which Katsonis was ordered to go to the Black Sea rowing fleet in command of Rear Admiral P.V. Pustoshkin in Odessa. However, Katsonis went to the Crimea, bought the area of Panas-Chir (Greek "Sacred Meadow") near Yalta, and began to build his manor, which he renamed as Livadia. He became a major industrialist, and builds a plant for the production of grape vodka. He was allegedly poisoned by a Turkish agent. George Gordon Byron took Katsonis as the prototype for his poem "Corsair." In Greece, their hero is highly honored, two submarines were named after him (in 1927 and in the 1980s). The life of this remarkable man was reflected in the story by Soviet writer V. Pikul "The First Listeragon of Balaclava."

Lazaretto Island

Lazaretto Island (*Lazareto*) – a small island with an altitude of 20 m. It is located 1.6 km southeast of the entrance to Govino Bay, which forms an inlet on the eastern coast of the island of Kerkyra, Ionian Sea, Greece, and covers the approach to it. A small pier was built at its western end. There is a small Church of the Savior and the ruins of Venetian buildings. There was a quarantine point in times of the British rule.

Lefkas

Lefkas, Leucadia, (*Lefkada*) (*from Greek "leukás" – white*) – island, in ancient times also known as Santa Maura, the fourth largest of the Ionian Islands. It is situated in close proximity to mainland Greece and between the islands of Corfu and Kefalonia. Lefkas is the only island that connects to mainland Greece by a long embankment road at the entrance from the Gulf of Corinth. Since the seventh century BC it has been inhabited by the Corinthians. It was once part of the mainland, but after in the sixth century BC, the Corinthians dug a canal 20 m wide, it was separated from it by a pontoon-articulated bridge 30 m long. From the south, the island is separated from Kefalonia by a wide (8.5 km) deep-water strait. The area of the island is 325 km^2. The population is 26,650 people (2011). The length of the coast is 117 km. The island is mountainous from east to west, divided by a series of mountains with the highest point of 1182 m covered with forest. There are pines, cypresses, many shrubs, olive groves, about 1 million trees, vineyards, fertile plains, and mountain creeks. Deep-water bays Rouda, Ŝivota, Afteli, and Vasiliki form an inlet into the high indented southern coast. At 2.2 km away from the middle part of the shore, there is Sesoula islet. The Vlykho bay forms an inlet into the high eastern coast – the most significant one along this coast (Fig. 2).

There are several islands and islets nearby. The largest island is Meganisi, separated from Lefkas by the navigable Meganisi strait. Lefkas and the islands located near it are surrounded by great depths. The main occupations of the inhabitants of the island are agriculture and tourism. The island is the largest producer of olives in Greece. In the northeastern tip of the island the city of Lefkada is situated. Shocks of earthquake are observed on the island. The island has been known since Homeric times. Over the centuries, the island passed from the Venetians to the French, from the Turks to the Russians, and then to the British, which left an imprint on the lifestyle and architectural appearance of the island. The east coast of the island is its most accessible and most developed part. A number of resort towns are located here: Nydri, Mikros Gialos, Sivota, and Vasiliki. The island is rich in long sandy beaches with white sand and beaches with fine flat pebbles. The most beautiful beaches are located on the west coast of the island. The most famous are Egremni, Porto Katsiki, Gialos, Kathisma, Milos, Agios Nikitas, Pefkoulia, Ai Giannis, Gira, Kastro, Yialos. Porto Katsiki is a fabulous beach with a snow-white sandy shore, located under a huge vertical cliff, overgrown with contrasting wild bushes and trees. It has the award for the best beach in Greece. Included in the top ten of the World's Beach Catalog (Fig. 3).

Fig. 2 Lefkas Island. (Source: https://adijuhpalace.ru/800/600/https/kuku.travel/wp-content/uploads/2017/11/ostrov-lefkada-3.jpg)

Fig. 3 Lefkas Beach. (Source: https://traveltimes.ru/wp-content/uploads/2021/06/scale_1200-1-29.jpg)

Egremni is an exotic beach with light blue sea, white cliffs overgrown with bright green bushes, and a milky sandy shore. Kathisma beach ("where the horizon converges") is 7 km of the purest fine sand, part of the beach is occupied by nudists. Lefkada has been considered an important strategic base for a long time. Even now you can see a series of fortresses here: the fortification chain ends with the Castle of Santa Maura (twelfth century). The windswept south of the island is a favorite spot for kitesurfing and paragliding fans. The island gained international fame thanks to the Greek magnate A. Onassis. His island Skorpios is situated near the coast of the island. In 2015, an earthquake with a magnitude of 5.2 occurred in the area of the island. The south-western part of Lefkas suffered. The island is connected by maritime traffic with the neighboring islands of the Ionian Sea, by plane – with Athens and Thessaloniki, by bus – with Athens and Patras.

(2) A group of 24 islands located between the islands of Kerkyra and Kefalonia.

(3) The main city of the island of the same name, located at its extreme northern end, fortified by a pier, directly opposite the mainland. The city is surrounded by water – from the west by the estuary, from the east – by the strait. The estuary is separated from the Ionian Sea by a sandy and pebble bay-bar with a 4 km long city beach. There are 4 of the 12 former mills preserved here. They bore Russian names – "Moskovskaya," "Orlovskaya," "Medzikovskaya," reminding that before the beginning of the twentieth century they grinded wheat grain supplied from Russia. The city was badly damaged by the earthquakes of 1948 and 1953, when only a few churches survived. After that, the islanders limited the height of the buildings by two floors. Most of the city is pedestrianized. Of interest is the Archaeological Museum, which contains an impressive collection of artifacts dating from the Paleozoic era to the fourth century AD, and one room is dedicated to the German archaeologist W. Dörpfeld, the phonograph museum, the historical museum of Lefkada and several churches – Agios Minas Cathedral, Evangelistria – the city's cathedral, Church of the Pantokrator, built in 1700 in gratitude for the liberation from the Turkish yoke in 1684 (restored after the earthquake).

(4) An artificial canal dug in the sixth century BC with a length approximately 5.5 km, 20 m wide, separating the island of Lefkada from the mainland in the western part.

Lefkatas

Lefkatas – (1) the most southwestern cape of the island of the same name. A steep cliff with a height of 75 m. Childe Harold from Byron's "Childe Harold's Pilgrimage" sailed past this cape, and contemplated "But when he saw the Evening star above Leucadia's far-projecting rock of woe, and hailed the last resort of fruitless love.... "This love meant the act of the ancient poetess Sappho, who jumped from unrequited love from a cliff on this cape. Today a jumping tournament is being held here.

(2) A lighthouse, built on the area of ancient ruins among amazing rocks in 1890 by the French during their short reign. On the southern tip of the island of Lefkada, which shoots out into the sea with a long narrow rocky promontory of 60 m above sea level, in ancient times there was a temple of Apollo. From the lighthouse you can see the two islands of the Ionian Sea – Kefalonia and Ithaca (Fig. 4).

Lepanto

Lepanto – a city in Greece (now – Nafpaktos). On October 7, 1571, during the Cyprus War of 1570–1573, a naval battle took place there between the fleets of Ottoman Empire (276 ships) and the "Holy League" (213 ships). Due to the best training, firearms, and protective equipment (the edged weapons prevailed in Turkish armament), the fleet of the "Holy League" was victorious. The Battle of Lepanto showed the increased importance of artillery in military operations at sea, and accelerated the development of

L

Fig. 4 Lefkatas Lighthouse. (Source: http://club-odysseys.com/mayak-ostrova-lefkada.jpg)

the sailing fleet. Ottoman Empire, taking advantage of disagreements among the allies, rebuilt its fleet and successfully ended the war by annexing the island of Cyprus.

Letojanni

Letojanni (*Letujanni*) – a cozy, very calm town located on the shores of the Ionian Sea and part of the large tourist area of Taormina, which is often called the Sicilian Monte Carlo, Italy. Letojanni is situated in the eastern part of Sicily, in the province of Messina, 170 km from the capital of the island – Palermo, 40 km southwest of Messina. The population is 2634 people (2004). The beaches are spacious, pebbly, with easy access to the water. The climate of the resort is subtropical Mediterranean. The temperature in January is +7–10 °C, in summer it can be quite humid and hot – up to +30 °C and higher. Among the main sightseeings of Letojanni is the neo-Gothic parish church of San Giuseppe, where you can see numerous frescoes and a beautiful depiction of The Pietà of the eighteenth century, a historical city center with monuments of local history and culture. On the main square of the resort there is a monument to the native of the city – the famous surgeon Francesco Durante (Fig. 5).

Lighthouse of Saint Theodoroi

Lighthouse of Saint Theodoroi (*Lighthouse of St. Teodoru*) – located 3 km from the town of Argostoli, on Cape Agios Theodoros. It is often called by the name of a cape. A unique building of its kind, the lighthouse is a sightseeing of Kefalonia island: 20 columns surround a circular Doric building, with a small turret at the top. The height of the lighthouse is 8 m. The building of the lighthouse began in 1828 by the British governor of Kefalonia, Charles Napier, and the was completed a year later. In his diary, he wrote the following: "The foundation of this small masterpiece of architecture was laid by Captain Kennedy on March 12 and 13, 1829" (Fig. 6).

According to legend, the lighthouse owes its shape to a small windmill nearby, which inspired the architects. In 1864, after the unification of the Ionian Islands with Greece, it was included in the

Fig. 5 Letojanni. (Source: https://pictures.novatours.eu/shared/uploads/ALL/Countries/Italy/Catania/Resorts/Letojanis_shutterstock_18634630.jpg?market=LVNOV&p=tab_content_wide)

general Greek lighthouse system. The earthquake of 1953 completely destroyed the originally constructed building. However, about 10 years later, in 1964, the lighthouse was restored in accordance with the original project thanks to the gratuitous plan of the architect Takis Pavlatos, under whose supervision the construction was carried out.

Lighthouse on the Cape Malagavi

Lighthouse on the Cape Malagavi – located in the northwest of Loutraki Bay. It was built in the nineteenth century. At that time this territory was actively used as a pier. However, the rock that rises here in the sea near the coast has caused the death of many ships. Therefore, a lighthouse was built here, where the Ambracian Gulf forms an inlet into the southern shore.

Lights of the Ionian and Adriatic Seas and Sicily Island

Lights of the Ionian and Adriatic Seas and Sicily Island – a maritime handbook, published in 2008 by the GUNiO of the Russian Ministry of Defense. It contains information about visual and sound-signal navigation equipment on the eastern and western shores and coastal waters of the Ionian Sea from Cape Tenaron to Cape Kefali and from Cape Santa Maria di Leuca to Cape Paci (Messina Strait), on the Ionian Islands, Sicily, Lipari Islands, and the island of Ustica, as well as on the shores and coastal waters of the Adriatic Sea. To the northwest of the line connecting Cape Mullet and Cape Santa Maria di Leuca. The description includes information about lighthouses, floating lighthouses, air lighthouses, illuminated signs, lights, floating lights, superbuoys-beacons (except the lights on a swinging base), and sound signal devices. Information about the

Fig. 6 Lighthouse of Saint Theodoroi. (Sourse: https://edam-media.costa.it/media_guid/zc11_89af994966be496281d14a5f8f119dea/original.jpg?Expires=1593521278&Signature=AwTlbCNgVzbhSdOelEAOjxUsWfDD1xrH7LZlP663yHIR1SVlZO9SD9rCsyi8DO9OiZFDDP0FR8WME7qEigY3dPYND3~zBYSPjSr8VnLZ31Sm9SiDgN5P-cEr~0eV6ZyantQng8ehCw1ndntyv13Siid5aqWbEDBUa0ZzJpGZQzQXJ3s6ZdW-e1vFGpj1UhC0cdHmHyqJYDXM7sMUrVxRuJ5fUCfWQxF2E~XsxkNSXwmZZTJ9FkkEzBE4jH9d7x2wTSib2mLJxGeSRA83Jslz3nzQLQv4qjZvNcZ9F45b-BoSkYt90lLi41wvzLchC-KLakbzgCfB9U2TU7bh1cbgUQ__&Key-Pair-Id=APKAJ47SWE2MWEJD256A)

luminous and non-luminous buoys, and landmarks, as well as about the lights on the drilling and oil rigs, mooring berths, and platforms are not included in the description.

Lignano Sabbiadoro

Lignano Sabbiadoro – the city is situated on a small peninsula between Trieste and Venice (it is 50 km away), between a broad delta of Tagliamento River, warm waters of the Adriatic Sea, and in the south by large drying Marano Lagoon, Friuli Venezia Giulia Region, Italy. The population of the city is 6800 people (2019). The first inhabitants came here in 1931 after nearby swamps were drained. There is a small harbor, a harbor complex for small vessels, designed for more than 5000 places, which is the largest in Italy and one of the largest in Europe. It is one of the main resorts in Northern Italy and the Adriatic coast. It was founded in the 1930s. The city has an 8 km area equipped with sandy beaches. In 1972 a sea terrace was built (Terrazza a Mare), where shops, bars, boutiques, solariums, etc. are located. The streets are filled with shops and supermarkets. There are Italian stone pine forests and green lawns. There are the tourist beach places near the city, such as Lignano – Riviera (Lignano Riviera), Lignano – Pineta (Lignano Pineta), and Lignano – Sabbnadoro (Lignano Sabbiadoro). There are the best children's camps in Italy, including the Adriatica children's health center, famous for its magnificent sandy beaches and an aquapark.

Likhud Brothers

Likhud Brothers, Ioannikios Leichoudes *(1633–1717)* and Sophronios Leichoudes *(1652–1730) (Brothers Likhud)* – Greek Orthodox monks, the first teachers of the

Fig. 7 Likhud Brothers. (Source: https://ic.pics.livejournal.com/maxbolshakov/71168348/186711/186711_900.jpg)

Slavic Greek Latin Academy – the first officially approved higher educational institution in the Russian state. Greeks by nationality, they came from Cephalonia, descendants of the Byzantine princely family of royal blood. Educated in Greece, then in Venice, and at the Padua University, they spent several years in Greece as teachers and preachers. When the Russian government planned to establish a higher school in Moscow with the name of the academy, Tsar Fyodor and Patriarch Joachim in 1682 asked the Eastern Patriarchs to send Orthodox and skilled teachers to Moscow. It so happened that from March 1683 the Likhud Brothers lived in Constantinople at the courtyard of the Jerusalem Patriarch Dositheos II. The latter invited Likhud Brothers to go to Moscow, where they arrived in March 1685 and in the same year began to teach at the Epiphany Monastery. In 1686, a special three-story stone building was built for the school in the Zaikonospassky monastery: the construction of the building was largely assisted by Prince Vasily Vasilyevich Golitsyn. All the students from the printing school were transferred to the new building to the first students of Likhud Brothers. All the students from the printing school were transferred, as well as "by the decree of the Tsars, up to 40 boyar children and a significant number of commoners." By the end of 1687, there were 76 students in the emerging Academy. In 1688, Ioannikios Leichoudes, with the rank of Russian ambassador, went to Venice, where he stayed for about 4 years. In 1694, at the request of Dositheus II, who by that time was very dissatisfied with the activities of the brothers (mainly due to conflicts of a selfish nature and denunciations), as well as the unseemly behavior of Ioannikios's son, Nicholas (he was granted a steward and owned significant land property), both brothers were removed from teaching at the academy and assigned for teaching at the Moscow printing house. In 1697, by decree of Peter I, they were instructed to teach 55 people the Italian language (out of 55, only 10 studied, the rest dissuaded). They continued to be accused either of heresies or of some kind of political intrigue in Constantinople (Fig. 7).

In 1704 they were exiled to the Hypatian Monastery in Kostroma. In 1706, Metropolitan Job of Novgorod arranged for the brothers to be transferred to Novgorod, where he instructed them to set up a Slavic Greek Latin school similar to that in Moscow. The Novgorod school of Likhud Brothers trained the first generation of Russian philologists. The school was located in Detinets, Novgorod, in a building that has since been called "Likhudov Corpus" (Pavilion of Likhud Brothers). Only two languages were left in the school for learning (that is why it was called "Greek-Slavic"). From Novgorod they were again summoned to Moscow for studies at the academy and to participate in the correction of the Bible, first by Sophronius in 1709, and after the death of Metropolitan Job – Ioannikios in 1716. Around 1720, after the death of Ioannikios, Sophronios was appointed abbot of Solotchin monastery a monastery in the Ryazan diocese. In 2007, a bronze monument was erected to them in front of the altar of the former Epiphany monastery in Moscow.

Limnoria

Limnoria – a marine crustacean woodworm that is widely spread in the Ionian Sea. It infests, destroys, as a rule, the outer layers of a tree: its passages are not deep (usually not deeper than 5 mm, very rarely – up to 15 mm from the surface of the tree), but sometimes it eats out hollows in the piles, the so-called "kettles." Limnoria, as a rule, cannot tolerate silty, stagnant, oxygen-poor water.

Lobsters

Lobsters – They belong to the family of sea decapod crustaceans. Their main habitat is rocky continental shelves. There are several types of lobsters, which vary considerably in appearance as well as in their taste characteristics. The average weight of European lobsters reaches 10 kg, and the length is slightly less than 1 m. According to their appearance, lobsters resemble ordinary crayfish in many ways. However, there is a significant difference between them, namely, chelate limbs of impressive size. Lobsters have a strong shell and ten legs, two of which have grown into chelae. Lobsters may be characterized by a variety of colors – from grayish-green to greenish-blue. There are red "antennas" on its body, and also a tail, in the form reminding a fan. Initially, lobster meat has a white color, but in the process of cooking it changes to red – due to this it is called the "sea cardinal." Very gentle taste and pleasant aroma are specific for lobster meat. The most delicious meat is found in the tail section of lobster. In its huge claws there is also meat, but it has a stiffer consistency. The green liver of lobster contained under the shell of the head, as well as its red caviar, is also of significant value. Lobster meat is considered one of the healthiest protein foods. It is low in calories, cholesterol, and fat than lean beef or chicken, but at the same time, it is rich in amino acids, potassium, magnesium, and vitamins B12, B6, B3, B2, and provitamin A. It is also a good source of calcium, iron, phosphorus, zinc, and sodium.

Loggerhead sea turtle (*Caretta-Caretta*)

Loggerhead sea turtle (*Caretta-Caretta*) – a species of sea turtles, the only representative of the genus *Caretta* (Loggerhead) or loggerhead sea turtles. An adult turtle reaches 70–95 cm and weighs 80–200 kg. Carapace is heart-shaped, brown, reddish-brown, or olive in color. The color of the plastron and skeletal bridge varies from cream to yellow. On the back there are five pairs of coastal scutes. Massive head covered with large scutes (Fig. 8).

The foreflippers are equipped by two claws. It feeds mainly on mollusks and crustaceans, as well as fish, jellyfish, and sponges. It is surprising that for millennia they have chosen the same coasts for nesting. Little turtles, hatching from eggs, can swim across the oceans, but return to breed on the same shore. The city authorities close the beaches at night for the nesting period – nothing should interfere with relict turtles. Turtle eggs used to be considered a delicacy – they were dug up and consumed on a large scale. Only when the number of turtles was significantly reduced because of this, a ban on turtle eggs was introduced everywhere. Females lay eggs at night 4–5 times per season, in summer and autumn. In a clutch there are 100–126 eggs with a leathery shell about 4 cm in diameter. Little turtles hatch after 1–2 months, lie motionless in the sand

Fig. 8 Loggerhead sea turtle. (Source: https://farm9.staticflickr.com/8326/8098098600_a48de2359b_b.jpg)

for about 26 more days, then dig out and go straight to the sea. Predators are ready for this – many turtle cubs never reach water. At night, the turtles go to the sea, focusing on the horizon – the faint light that it radiates. If they are disoriented at this moment, they will confuse the direction and die. Having reached the sea, they can feel safe – they have few enemies in the water. Loggerheads are common in the Indian, Pacific, and Atlantic Oceans (including the Mediterranean Sea). A large batch of loggerheads nest on the island of Zakynthos. Loggerheads are listed in the IUCN Red List of Vulnerable Species, in the list of the Convention on International Trade in Endangered Species of Wild Fauna and Flora. Protected by the national laws of Greece, Italy, Cyprus, Turkey, and the USA.

Lokris

Lokris – a small region of Central Greece, in ancient times inhabited by the Locrians. Its western part or Ozolian Locris was a narrow strip in the north of the Gulf of Corinth. From the west it bordered on Aetolia, and from the east – on Phocis. From 338 BC, it was a member of the Aetolian Union, and regained its independence after the victory of the Romans (167 BC). Its eastern part was located on the shores of the Euboean Strait of the Aegean Sea.

Long Walls

Long Walls – in ancient Greece, fortifications that connected a city located at a distance from the sea with its port, or ports. Thanks to this, the city-dwellers were able to withstand prolonged sieges from land, being able to supply food and other goods by sea. Corinth was connected by Long Walls with Lechaeum (or Lechaio), Athens – with Piraeus and Phaleron. Megarians, with the help of the Athenians, built walls between Megara and Nisea, the Argives between Argos and Temeni. There were also own “Long Walls” in Patras and Milas (Mylasa). The most famous “Long Walls” were the Athenian ones.

Love Channel

Love Channel (*Canal d’Amour*) – romantic place, a bay, located next to Cape Drastis in the area of the village of Sidari in the extreme northwest of the island of Kerkyra, Greece. The channel runs between sheer sandstone cliffs. Its walls are eroded, and the winds have adorned the stones with ornate patterns extending into the Ionian Sea, forming a narrow body of water that resembles a canal. According to legend, these rocks used to be "lovers," causing the envy of the gods with their happiness, and therefore were turned into stone rocks. It is believed that if the lovers float along this channel, then nothing in the world would be able to separate them. It is also believed that if a woman swims past two rocks and guesses the name of her beloved, then he will definitely marry her.

M

Maddalena

Maddalena – peninsula on the east coast of the island of Sicily, Italy. It is composed of steep rocky cliffs 15 m high. Behind them, there are hills, ascending to a height of 50 m, the slopes of which are cultivated.

Madouri

Madouri (*Maduri*) – an islet 44 m high, located in the middle of the Meganisi Strait east of Lefkada Island and west-southwest of the southern tip of Sparti Island, Greece. It belongs to the family of the famous Greek poet Valaoritis.

Magna Graecia

Magna Graecia – the name of main Greece accepted in ancient Rome.

Mahi-mahi

Mahi-mahi – common dolphinfish (*Coryphaena hippurus*) – fish of the spar family Sparidae. Predator. Length up to 1 m. The maximum weight is 39 kg. One of the fastest-growing fish in the sea. The sides are flattened, and the head is blunt, large. Long dorsal fin and sharply concaved caudal fin. The color is ash-silver with a golden hue, which quickly fades after death, becoming completely yellow. It is found in the Mediterranean seas. From the time of Ancient Rome to the nineteenth century, it was an object of aquaculture. It migrates over the long distances. Medium quality meat. It has a significant protein content, which makes it quite nutritious. Fish jump out of the water and often changes color. Known to the world for a long time, and has always been the most popular fish in the Mediterranean. In ancient Rome, the meat of this fish was highly valued, thus it was specially kept and fattened in salted ponds and pools. In Venice, this method of fattening existed at the end of the nineteenth century. It feeds on crabs, squid, mackerel, and flying fish. Artificial cultivation of this delicacy fish has been mastered since very ancient times and is now widely spread in France, Italy, Turkey, Greece, and other countries. To obtain tender meat and excellent taste, it is bred in exclusive conditions: in closed rooms with a special light that imitates a certain season. Catches in the Mediterranean reach approximately 800 t.

Maitland Thomas (1759–1824)

Maitland Thomas (*1759–1824*) – sir, lieutenant general, British officer, and colonial governor.

I. S. Zonn, *The Ionian Sea Encyclopedia*, Encyclopedia of Seas,
https://doi.org/10.1007/978-3-031-08206-1

Member of parliament. Member of the Privy Council of Great Britain. Maitland was enlisted in the Edinburgh Light Cavalry Regiment shortly after his birth, but when he came of age, he joined the 78th Scottish Infantry Regiment with the rank of captain (1778). In 1790 he was transferred to the 72nd Infantry Regiment, and then to the 62nd Infantry Regiment with the rank of Major. Finally, in 1794, he was promoted to colonel general, and in 1798 brigadier general. In 1805–1811. served as Governor of Ceylon (Sri Lanka). In 1813, Maitland was appointed Governor of Malta after it became a British colony, and remained so until his death in 1824. When he was Governor of Malta, Maitland also served as Lord Commander of the Ionian Islands, which became a British protectorate, he was the first British High Commissioner (ruler from the mother country) of Corfu and the Ionian Islands (Fig. 1).

Fig. 1 Maitland Thomas (*1759–1824*). (Source: http://2.bp.blogspot.com/-pfwO5p8MU7w/Wsb7GuUXbdI/AAAAAAAAhRE/I7FkbIyLClMFxbO58-aTMOZW9ff9-aI8QCK4BGAYYCw/s1600/%25D0%25BC%25D1%258D%25D0%25B9%25D1%2582%25D0%25BB%25D0%25B0%25D0%25BD%25D0%25B4%2BLieutenant%2BGeneral%2BSir%2BThomas%2BMaitland%2B%25281759%25E2%2580%25931824%2529%252C%2BGovernor%2Bof%2BMalta%2B%25281813%25E2%2580%25931824%2529.jpg)

Makronisos

Makronisos – Makri (*Nissos Makri*) (*Makronisos*) – an island that is part of the Echinades Islands, Greece. Located 2.2 km east of the island of Vromonas Ionian Sea. In the northern part of the island there is a cone-shaped hill 127 m high. The shores are deep, except for its southeastern and northwestern ends, bordered by shoals with rocks (Fig. 2).

Mani

Mani – the southernmost peninsula in Greece, one of the three tips, "finger tips" of the larger Peloponnese peninsula. Mani is located in the middle, being the middle "finger." It covers the territory from Gytheio in the east to Kardamili in the west and ends in the south with Cape Tainaron, which serves as the point of separation of the Ionian Sea from the Aegean Sea. The western part of Mani is washed by the Ionian Sea, and the eastern by the Aegean, and according to myths, there was an entrance to the underworld here. The landscape is wild and the climate is dry. The peninsula itself consists of two parts that are not alike: Exo Mani (Outer Mani) and Meso Mani (Inner Mani), separated by a gorge near Itilo. Inner Mani occupies the area south of the line connecting the bays of Itilo and Vathi – this is actually the real Mani: a ragged coastline, sparse indentations, and rocky land. Outer Mani is part of the Ionian Sea coast from Areopoli to Kalamata, part of Messinia. Its nature is rich and diverse. In the vicinity of Kardamili there are the Viros and Stoupa gorges – these are the best resorts. Mount Taygetus is one of the highest mountains of the Peloponnese (2408 m), which is popular among climbers and tourists. Its slopes go out to the Ionian Sea. The Mani Peninsula has a long and turbulent history. Ancient Mani was controlled by the ancient Spartans, who were infamous outside the region for their overly harsh temper and military order. The Christianization of the region somewhat softened the local mores, but could not destroy the values of a deeply patriarchal society, where personal ties stood above the laws, and blood feud became

Fig. 2 Makronisos, Makri. (Source: https://cdn.pixabay.com/photo/2018/02/22/23/50/cyprus-3174317_1280.jpg)

M

commonplace. Medieval Mani became famous not only for the endless bloody feuds between the local clans of the Maniots, but also for the rivalry of various states (Byzantium, Venice, the Ottoman Empire) with each other, as a result of which many beautiful castles appeared on the peninsula. Since the sixteenth century rival clans used the fortress towers to bombard opponents. By the end of the eighteenth century a gradual unification of clans begins to fight for the independence of Greece against the Ottoman Empire, which flared up in 1821–1830. Many Maniots also participated in mass pogroms against local Muslims and sought to ensure complete homogeneity of the population on the peninsula in terms of its ethnic and cultural composition. Yet strife and violence continued here until the end of the XIX century.

Maquis

Maquis – thickets of evergreen stiff-leaved and thorn bushes, short trees, and tall grasses in arid subtropical regions. The most common on the slopes of mountains and hills in the Mediterranean climate, especially in the continental Balkan regions and in other regions of the world with a similar climate, local terms are used to designate them – scrab (Australia), chaparral (North America), in South Africa finbosch (Cape Region). Maquis is extremely flammable. A clear confirmation of this was the fires in Montenegro (2011). However, fire for Maquis is a natural source of vegetation rejuvenation and enrichment of the soil with minerals necessary for the germination of new seeds. Different types of maquis are the most typical in the modern Mediterranean countries, where they replaced the forests, cut down by men in antiquity times, although there are also indigenous massifs. Maquis are the most specific for the lower zone of the mountains up to a height of 800 m, where they form dense, thorny, and impassable thickets. Maquis thickets are presented by the two levels. The first level is the maquis itself, which consists of short growing trees, growing at a height from 0 to 400 m above the sea level. Further, shrubs and grasses constitute the garigi strip. Maquis

Fig. 3 Maquis. (Source: https://i.pinimg.com/originals/93/18/cb/9318cbb6668133d598bd04998aa4b3fc.jpg)

includes a large number of species. However, thorny bushes predominate (average height is 2–4 m), and trees of 8–10 m height are less common. There are the following species: pistachio, arbute, juniper, olive, rockroses, myrtle, etc. Grass cover is dominated by annual grasses. Many of them have an increased concentration of ethereal oils, and therefore have a strong odor (Fig. 3).

Tall and dense thickets of xerophytic shrubs (madrone, myrtle, rockrose, tree heath, Mt. Atlas mastic tree, olive, rosemary, junipers). Maquis degenerates into sparse and scarce plant groups known as "garrigue."

Marathonisi

Marathonisi – or "turtle" island 148 m high, length from north-west to south-east approximately – 925 m, width – 185, overgrown with olive trees, lies on the southeastern end of the rocky spit, extending from the western coast of Zakynthos, Keri Bay, Greece, 3.3 km to the south-east. There is an old tower at the top of the island. Marathonisi is divided into two parts: tourist and turtles, where the Caretta Caretta turtles lay their eggs, where there are hundreds of them. Tourists are allowed to visit this island, but only before sunset.

Mariculture

Mariculture – a commercial cultivation of marine or brackish-water organisms on natural or artificial feed in fenced-off gulfs or special cages. The main volumes of mariculture are algae, clams, and shrimps. Fish farming in mariculture has a subordinate meaning. Sometimes it is called sea farming. A mariculture specialization was developed in Europe, which consisted mainly of two groups of mollusks – oysters and mussels. The most common

Fig. 4 Mariculture. (Source: https://sustainablefisheries-uw.org/wp-content/uploads/2018/08/shutterstock_789025267.jpg)

marine species grown in the Ionian Sea, as well as in the Mediterranean Sea, are euryhaline species: European bass (*Dicentrarchus labrax*) and gilt-head sea bream (*Sparus auratus*). For the start of their breeding, juveniles of different sizes are used, usually with an average weight of 2–4 g (age 120–160 days). Breeding of Atlantic bluefin tuna (*Thunnus thynnus*) began in mid-1990s. Recently, "new species" for nurseries culture have emerged: Sharpsnout seabream (*Diplodus puntazzo*) and Meagre (*Argyrosomus regius*), as well as common dentex (*Dentex dentex*), Red porgy (*Pagrus pagrus*), Common two-banded seabream (*Diplodus vulgaris*), etc. Their further introduction depends on the reaction of the market, which is still accustomed to traditional bred species. However, the number of protected areas along the coastline is limited, which prevents the expansion of nursery farms. In addition, tourism (the main source of income) is often a competitor in the use of marine and coastal resources. About 60% of marine nursery farms are currently located in semi-offshore or offshore zones, which entails higher production costs and the need to deal with various technological issues related to nurseries models, positioning, and installation (REFA Tension Legs, steel nurseries Sadco Shelf, Farmocean and a number of submerged nurseries, floating fish platforms). Several Atlantic bluefin tuna feed farms have recently emerged (Fig. 4).

Marie Henri Daniel Gauthier

Marie Henri Daniel Gauthier – comte de Rigny (1782–1835) (*Marie Henri Daniel Gauthier de Rigny*) – Count, French statesman and military leader, admiral, and diplomat. Born in the French city of Toul. At the age of 15, in 1797 he entered the naval service. In 1809 he became a lieutenant, in 1816 – a captain. In 1822–1825, chief of the French squadron in the Levant. From 1825 – Rear Admiral. He commanded a French squadron in the Battle of Navarino, after which he became vice admiral. French Naval Minister in 1831–1834 Minister of Foreign Affairs of France from April 4 to November 10, 1834, and from November 18, 1834, to March 12, 1835. In honor of the 100th anniversary of the Battle of Navarino in Greece in 1927, a postage stamp was issued with the image of de Rigny.

Marina

Marina – harbor and base, equipped parking pier for pleasure and sports boats (yachts and boats) in Croatia, Montenegro, and other countries. On large Marinas there are all the necessary conditions for the maintenance of yachts, as well as

Fig. 5 Marina. (Source: https://www.charterworld.com/images/headers-2/Port%20Vauban%20Antibes.jpg)

shops and restaurants. Yacht charter system is developed. If necessary, you can get training and instruction. In Croatia, there is the Association of Independent Marines, where you can get information about the conditions of yachts and boats charter (Fig. 5).

Marine Aquaculture

Marine Aquaculture – see Mariculture.

Marine spatial planning (MSP)

Marine spatial planning (MSP) – is the main instrument of Integrated Coastal Zone Management (ICZM). Its essence lies in inventorying and planning the use of marine areas, spatial analysis of potential conflicts between man and nature, between different users of natural resources, analyzing the needs of current and future use of marine areas, developing environmentally friendly scenarios for using marine areas, and developing plans for long-term comprehensive and sustainable economic use of marine areas. Maritime planning is based on the following principles: intersectoral coordination and cooperation of all the parties concerned, coordination of development strategies and environmental management of marine areas developed at the national, regional, and local levels, creation of a comprehensive information base, extensive use of geographic information systems (GIS), including information on economic and socioeconomic activities, the state of the environment, the creation of instructions for the maritime users on the sustainable use of marine resources, interaction and cooperation of citizens and organizations within and outside the region, as well as an effective system of cross-border consultations.

Maritime – Folklore Museum of Ithaca

Maritime – Folklore Museum of Ithaca – is located on the island of Ithaca. There is also the Museum of Folklore under one roof with it.

Maritime and Historical Museum

Maritime and Historical Museum (Museum of Navigation and History) – opened in 2005 in Galaxidi, one of the former main Greek harbors on the northern coast of the Gulf of Corinth in Thessaly and Central Greece. The exposition presents the settlements that existed on the site of Galaxidi over a long history – these are ancient Halleion and Oeantheia, then settlements from the Byzantine period to our times. During its prosperity, the city had half a dozen shipyards. Here you can trace the biography of two- and three-masted ships from their creation to their death. Logs, wooden rattles were collected, announcing, instead of watch-glass, a change of watch, a boon – a large shell, the sound of which served as a signal in fog and bad weather, painted sculptures installed on the forward bows of ships.

Marrobio

Marrobio – a strong and sudden surge of water noted off the coast of the island of Sicily and nearby islands. The water recedes from the coast at the same speed, carrying silt, and algae with it.

Mathraki

Mathraki (*Samothrace*) (*Matrak*) – the island is a satellite of the Kerkyra island. Located to the northwest, 8 km from the Ionian Sea island of Kerkyra and 4.6 km west of the Diaplo island. The length of the island is 3600 m, and the greatest width in the southernmost part is 1200 m. The area is 3.1 km^2. The population is 329 people (2011). The island is bordered by a shoal, on which surface and underwater rocks are scattered. It is rough, mountainous, with a maximum height of 152 m in the middle. Covered with forest. The 3-kilometer-long Portello Beach, composed of dark red sand, stands out (Fig. 6).

Fig. 6 Mathraki Island. (Source: https://www.esys.org/rev_info/Griechenland/Erikoussa_Luftbild_Insel.jpg)

Mediterranean Basin

Mediterranean Basin – Tyrrhenian, Adriatic, Ionian, Aegean, and the Mediterranean Sea itself. The Marmara, Black and Azov Seas, separated from the Mediterranean Sea and from each other by the straits, enter the basin of the Mediterranean Sea. The internal seas, which are separated from each other by islands, are also included into the Mediterranean Sea basin. Sometimes these seas are divided into even smaller parts: the Alboranian Sea – the narrow part of the Northern Sea near the Strait of Gibraltar, and the Balearic and Sardinian Seas – the area between the coast of Spain and Sardinia Island. The Ligurian Sea is the northern part of the Tyrrhenian Sea, the Cretan Sea is part of the Aegean Sea at the Crete Island, and the Levantine Sea is the part of the Mediterranean Sea between Turkey and Africa.

Mediterranean Expedition of Black Sea Navy

Mediterranean Expedition of Black Sea Navy – Mediterranean Campaign of F.F. Ushakov 1798–1800 – combat operations of the united Russian-Turkish squadron (27 ships) under command of Vice-Admiral F.F. Ushakov, with the aim of eliminating French dominion in the Adriatic Sea and in the eastern part of the Mediterranean Sea (Ionian Sea) and assisting the allied forces in liberating Italian territory from the French during the War of the Second Coalition against France. From September 1798 to August 1799 – the Russian-Turkish squadron captured a number of the Ionian Islands, including Corfu, assisted in the offensive of the allied forces (Field-Marshal-General – A.V. Suvorov) in Italy, took part in joint actions with the British squadron (Rear Admiral – G. Nelson). As a result of the mutiny of the crews of the ships, the Turkish squadron (8 ships, Vice Admiral Kadir Bey) unauthorizedly left for the Dardanelles. The Russian squadron continued to carry out combat missions until June 1800 and returned to Sevastopol in October. The expedition set an important precedent in Russian and world history. This is a bright page of Russian naval glory, which had military and geopolitical significance. The issue on the status and form of government for the Ionian Islands was resolved. The Septinsular Republic was established with one of the most democratic constitutions of that time.

Mediterranean monk seal

Mediterranean Monk Seal (Monk Seal, *Monachus monachus*) – like all pinnipeds (Pinnipedia), it is found in water and on land, near the seashore. The length of the seal reaches 3.5 m. Weight is up to 250 kg. Develops speed from 15 to 25 km/h. It can dive to a depth of 300–350 m, and being beneath the waves for 25–30 min. The color of the dorsal part of the body is gray-brown, and the ventricular part is whitish. Mediterranean monk seal prefers uninhabited rocky coasts, with numerous caves, where animals breed. The female, after 10–11 months of pregnancy, probably gives birth to one puppy at the beginning of spring, which feeds on milk for 3–4 months, then begins an independent life. Mediterranean monk seal lives for 30–40 years. It feeds mainly on fish, preferring individuals of larger species. Mediterranean monk seal disappeared in most previously known habitats. It is so rare in the Adriatic Sea that there is a serious danger of its complete disappearance. It is listed in the international "Red Book." (Fig. 7).

Mediterranean Moray (*Muraena helena*)

Mediterranean Moray (*Muraena helena*) – large serpentine fish of Muraenidae family. The closest relative of other serpentine fish – eels. All of the moray species are characterized by large sizes: the smallest ones reach a length of 60 cm and weigh 8–10 kg, while the world's largest giant moray (*Thyrsoidea macrura*) reaches a length of 3.75 m and weighs up to 40 kg! Moray's body is

Fig. 7 Mediterranean monk seal. (Source: https://cherepah.ru/wp-content/uploads/d/9/d/d9d8580bf3c22a0f5cbfc8caf04cf101.jpeg)

disproportionately long, slightly flattened on its sides, but not completely flat. The back of the body looks thinner, while the middle and front of the body are slightly thickened. Due to that moray resembles a giant leech. The pectoral fins of these fish are totally absent; however, the dorsal fin stretches along the entire length of the body. Few people manage to see moray in all its glory. In most cases its body is hidden in the crevices of the rocks, and only the head sticks out. The snout of moray is elongated with an angry expression of its eyes, the mouth is almost always open, and large, sharp teeth can be seen in it. This was the reason to accuse morays of serpentine cunning and aggression. In fact, the expression of the eyes of moray is not so much vicious as frozen, because these are ambush fishes, who spend a lot of time waiting for prey. Moray often sits with an open mouth, because they breathe through it, since in tight shelters the flow of water to the gills is difficult. Moray has no tongue; however, it has two pairs of nostrils. Like all fish, they use their nostrils not for breathing, but only for smelling. The color of these fish is variegated, most often with a finely topped pattern (less often striped, monophonic), while colors are usually nondescript – brown, black, and whitish-gray. These fish are mostly found at shallow depths of up to 5 m; some species can creep over land at low tide. These specificities make morays very similar to their relatives, eels. Morays keep most of their lives in shelters: crevices of underwater rocks, internal cavities of large sponges. These fish are active mainly at dusk, therefore they have a bad sight, but they compensate for this deficiency with an Morays live alone and adhere to permanent areas. In rare cases, when several convenient crevices are found nearby, morays may live side by side with each other, but this is a random neighborhood. Moray's bites are not only very painful but also extremely poorly healed (up to several months), and fatal causalities are also known. Moray is a predator. It feeds on fish, crabs, sea urchins, octopuses, and cuttlefish. It lurks its prey in ambush. When approaching the victim at a sufficient distance, moray, with a flash-light hurl, throws the front of the body forward and grabs the victim. Moray larvae are called leptocephalus. They have a rounded head and a rounded tail fin, their body is completely transparent, and the length at birth barely reaches 7–10 mm. It is extremely difficult to see such a larva in water, and leptocells also swim freely and are carried by currents over fairly long distances. Thus, sedentary morays spread. The drift period lasts for 6–10 months, during which time the leptocephalus grows and begins to lead a sedentary lifestyle. Morays reach sexual maturity at 4–6 years. The lifespan of these fish is not exactly established, but it is great. It is reliably known that most species can live for more than 10 years. There are complex relationships between man and morays. On the one hand, people always feared these predators and avoided close contact with them in their natural environment. On the other hand, morays meat has been famous for its excellent taste since ancient times. Famous gourmets, the ancient Romans, valued Mediterranean Moray meat on a par with the meat of its freshwater and smaller relative – eel. Morays are served at feasts as a delicacy and in large quantities. Therefore, despite the fear, people from ancient times caught morays, and the Romans even learned to breed them in cages. Now the experience of breeding morays in captivity is lost. It tolerates imprisonment well and often lives in aquariums (Fig. 8).

Mediterranean Sand Smelt (*Atherina hepsetus*)

Mediterranean Sand Smelt (*Atherina hepsetus*) – marine schooling pelagic fish of the Sand smelts family (Atherinidae). The body is fusiform, cylindrical, flattened on the sides. The head is small. Mouth and eyes are big. The teeth are small. The tail is branched. Color is dark green with black small spots. The sides and belly are silvery white. Along the sides from the head to the tail a silver stripe stretches. Length up to 20 cm. Spawning is portional, and occurs in the spring in the coastal zone. It is widely spread throughout the Adriatic Sea. It usually occurs at depths from 5 to 20 m. The life expectancy is 3–4 years. Meat is delicious.

Mediterranean Science Commission

Mediterranean Science Commission, (*Commission Internationale pour Exploration Scientifique de la Mer Méditerranée*) (*CIESM*) – in 1908, at the IX International Congress of Geographers, a special committee was created, headed by Prince Albert I of Monaco, to study the Mediterranean Sea. At its first meeting in 1910, the committee became known as the International Commission for Scientific Exploration of the Mediterranean (ICSEM). Regular meetings of this Commission began only after the 1919 General Assembly in Madrid, at which its main goals and objectives were formulated. ICSEM was created with the aim of cooperation of the Mediterranean countries for the coordination and planning of exploration and the timely exchange of scientific information in the field of oceanology, research and pollution

Fig. 8 Mediterranean Moray. (Source: http://komotoz.ru/photo/zhivotnye/images/photo_mureny/photo_mureny_01.jpg)

control, and research into the problems of human penetration into the depths of the sea. ICSEM is an intergovernmental consultative organization composed of representatives of the contracting states. The President and Chief Officer are elected for 4 years. Its main bodies are the Plenary Congress Assembly, the Board of Governors, the Advisory Committee, and the Scientific Council. The Commission has a charter and a budget. The Commission has 23 member states: Algeria, Croatia, Cyprus, Egypt, France, Germany, Greece, Israel, Italy, Lebanon, Malta, Monaco, Morocco, Portugal, Romania, Russia (since 2013), Slovenia, Spain, Switzerland, Syria, Tunisia, Turkey, and Ukraine. Practical activities are carried out in scientific committees: marine geosciences, ocean physics and climate, marine biogeochemistry, marine microbiology and biotechnology, living resources and marine ecosystems, and coastal ecosystems. About 2000 scientists from more than 30 countries take part in the work of these committees. They conduct seminars and symposia that focus on emerging scientific issues. From its inception to 2014, the Commission held 40 congresses, five of which took place in the cities of the Adriatic: Venice (1926, 2010), Split (1976), Trieste (1992), Dubrovnik (1998). From 1920 to 1926, the Commission published the Bulletin ("Bulletin de la Commission Internationale pour I'Exploration Scientifique de la Mer Méditerranée"), and then it was replaced by the series "Reports of the Congress" ("Rapports du Congrès"), which is published and now. The official language of the Commission is French. In its work, the Commission is associated with the IAEA, FAO, IOC, UNESCO, ICSU, etc.

Mediterranean Sea

Mediterranean Sea – a sea in the Atlantic Ocean. It is the Mediterranean Sea, which means that it is located between Europe, Africa, and Asia. In the west it connects with the Atlantic Ocean through the Strait of Gibraltar, in the northeast through the Dardanelles with the Sea of Marmara, then through the Bosporus with the Black Sea, in the south-east through the Suez Canal with the Red Sea and then the Indian Ocean. The area of the Mediterranean Sea is 2.5 million km^2, water volume is 3.84 million km^3, average depth is 1498 m, and the maximum depth is 5121 m. Seas of the Mediterranean Sea basin wash the shores of 21 states of three continents – Europe, Asia, and Africa: Tunisia, Algeria, Morocco, Egypt, Libya, Lebanon, Israel, Italy, Spain, France, Greece, Cyprus, Turkey, Syria, Monaco, Malta, Slovenia, Croatia, Montenegro, Albania, and Bosnia and Herzegovina. The name (Greek, ΜεσόγειοΒ Θάλασσα; Lat. Mare Mediterranea – the sea among the lands) was first introduced by the ancient writer Gaius Julius Solinus, who based on the beliefs of his times, as ancient European and North African civilizations were developed in the basin of this sea, which was the only traffic connection between them. The following seas are parts of the Mediterranean: the Alboran Sea, the Balearic Sea, the Ligurian Sea, the Tyrrhenian Sea, the Ionian Sea, and the Aegean Sea. The Mediterranean Sea basin includes the Marmara, Black, and Azov Seas. The shores of the Mediterranean Sea at the mountainous coasts, predominantly abrasive, leveled. However, at the low ones, they are lagoon-liman and deltoid; for the eastern coast of the Adriatic Sea typical shores of the Dalmatian type. The most significant gulfs are: Valencia, Lyon, Genoa, Taranto, Sidra (Syrtis Major). The largest islands are Balearic Islands, Corsica, Sardinia, Sicily, Crete, and Cyprus. The large rivers, such as Ebro, Rhone, Tiber, Po, Nile, and others, empty into the Mediterranean Sea. Hydrological regime of the Mediterranean Sea was formed under the impact of large evaporation and general climatic conditions. The predominance of freshwater expenditure above the inflow leads to a decrease in the sea level, which is the cause of the constant inflow of surface, less saline waters from the Atlantic Ocean and the Black Sea. In the deep layers of the straits, there is an outflow of high-saline water caused by the difference in the water density at the level of the thresholds of the straits. The main water exchange takes place through the Strait of Gibraltar (the upper course brings 42,320 km^3

per year of Atlantic water, and the lower reaches 40,800 km^3 of the Mediterranean); through the Dardanelles, 350 and 180 km^3 of water per a year flow in and out, respectively. The circulation of waters in the Mediterranean Sea has mainly wind-driven nature. It is represented by the main, almost zonal current carrying waters of predominantly Atlantic origin along Africa, from the Strait of Gibraltar to the shores of Lebanon, a system of cyclonic gyres in isolated seas and basins. The water column to a depth of 750–1000 m is covered by unidirectional water transfer along the depth, except the Levantine intermediate backflow, which transfers the Levantine waters from the eastern Mediterranean to the Strait of Gibraltar. The rates of steady currents in the open sea are 15–30 cm/s, in some straits 55–110 cm/s. The average temperature of water on the surface in February decreases from north to south from 8–12 °C to 17 °C in the eastern and central parts and from 11 °C to 15 °C in the west. In August, the average water temperature varies from 19 °C to 25 °C, and in the extreme east it rises to 27–30 °C. Large evaporation leads to a strong increase in salinity. Its values increase from west to east from 36 to 39.5‰. The density of water on the surface varies from 1.023 to 1.027 g/cm^3 in summer and to 1.027–1.029 g/cm^3 in winter. During the period of winter cooling, intensive convective mixing develops in areas with a higher density, which leads to the forming of highly salinized and warm intermediate waters in the Eastern Basin and deep waters in the north of the Western Basin, in the Adriatic and the Aegean Seas. According to the bottom temperature and salinity, the Mediterranean Sea is one of the warmest and saltiest seas of the World Ocean (12.6–13.4 °C and 38.4–38.7‰, respectively). The relative transparency of the water reaches 50–60 m, and the color is intensely blue. The tides are mostly semi-diurnal, their magnitude is less than 1 m; however, at certain points, in combination with wind surges, the level fluctuations can reach 4 m (Gulf of Genoa, off the northern coast of the Island of Corsica, etc.). In the narrow straits there are strong tidal currents (Strait of Messina). The maximum heave of the sea is noted in winter (the height of the waves reaches 6–8 m). The vegetation and animals of the Mediterranean Sea are characterized by a relatively poor quantitative development of phyto- and zooplankton Mediterranean Sea is characterized by a relatively poor quantitative development of phyto- and zooplankton, which entails a relatively small number of larger animals that feed on them, including fish. The amount of phytoplankton in the surface horizons is only 8–10 mg/m^3, and it is 10–20 times less at a depth of 1000–2000 m. Algae are very diverse, and peridinean and diatoms predominate. The fauna of the Mediterranean Sea is characterized by a large species diversity; however, the number of individual species representatives is small. There are crayfish, one species of seals (Mediterranean monk seal), and sea turtles. There are 550 species of fish: mackerel, herring, anchovy, mullet, grenadier, tuna, bonito, horse mackerel, etc. There are about 70 species of endemic fish, including skates, species of anchovies, gobies, blennies, wrasse, and garfish. Among the edible mollusks, there are oysters, Mediterranean Black Sea mussel, and date mussels are of the greatest importance. Invertebrates are presented by octopus, squid, sepia, crabs, and lobster are common; there are numerous types of jellyfish and siphonophores, and in some areas, especially in the Aegean Sea, sponges and red coral dwell. The shores of the Mediterranean Sea have long been densely populated, and have a high level of economic development (especially the nations, situated on its northern coast). The agriculture of the Mediterranean countries is specified by the production of citrus fruits (about one-third of the world harvest), cotton, and oil-bearing crops. In the system of international trade and economic relations, the Mediterranean Sea occupies a special position. Being at the junction of three parts of the world (Europe, Asia, and Africa), the Mediterranean Sea is an important transportation route, through which maritime connections between Europe and Asia, Africa, North and South America, as well as Australia and Oceania pass. Important trade routes connecting Russia and Ukraine with Western countries pass through the Mediterranean Sea. The transport value of the Mediterranean Sea area for Western Europe is continuously increasing due to the increasing dependence of these countries on the import of raw materials. The Mediterranean Sea is an important "oil" route between Western Europe and the Middle East. The share of southern ports (the main of which are Marseille, Trieste, and Genoa) in the supply of oil to Western Europe is constantly growing. The transportation of various types of raw materials, metal ores and bauxite, and agricultural products along the Suez Canal, through which the connections of Western Europe with Asia and Australia pass, is also great. The largest ports are Marseille with avant-ports in France; Genoa, Augusta, Trieste in Italy; Sidra, Marsa-Brega in Libya. Numerous industrial enterprises are created on the coast of the Mediterranean Sea and on the islands. Oil and gas production on the shelf of the Mediterranean Sea (northern part of the Adriatic Sea, coast of Greece, etc.) has begun. Compared with other basins of the Atlantic Ocean, fisheries in the Mediterranean Sea are of secondary importance. The development of industry on the coast and the growth of cities, seaports, and recreational areas leads to intensive pollution of the coastal strip. The resorts of the Cote d'Azur (Riviera) in France and Italy, the resorts of the Levant coast, the Balearic Islands and coast in Spain, and the resorts in the Adriatic and Aegean seas are very famous (Fig. 9).

Mediterranean Seas

Mediterranean Seas – seas, extremely extending into the land and connecting to the ocean through one or more straits. They are divided into intracontinental and intercontinental, depending on whether they are surrounded by the land of one continent (the Baltic Sea) or located between two continents (the Mediterranean and the Red Sea). Compared to other seas, they are characterized by the greatest isolation and, therefore, the greatest specificity of the hydrological regime, which is greatly affected by the surrounding land.

Meganisi

Meganisi – (1) island (Greek – large island), located to the east of the island of Lefkada, separated from it by the Meganisi Strait of the Ionian Sea, Greece. Area is 20 km^2, and the population is 1041 people. (2011). Tourist infrastructure is almost absent. The place is popular with yachtsmen. There are three villages (townships) and two ports on the island. The island is distinguished by a unique variety of coves, sandy and pebbly beaches, an abundance of caves, traditional villages with narrow cobbled streets, beautiful

Fig. 9 The Map of the Mediterranean. (Source: https://upload.wikimedia.org/wikipedia/commons/thumb/f/f3/Map_of_the_Mediterranean_and_subdivisions.svg/2560px-Map_of_the_Mediterranean_and_subdivisions.svg.png) Gabes (Syrtis Minor)

M

Fig. 10 Meganisi Island. (Source: https://asteria.gr/wp-content/uploads/2017/08/Asteria-Boats-Villas-Meganisi-Island-2-1536x864.jpg)

landscapes of private islands and Lefkada, and hospitable and incredibly cheerful people. In the northern part of the island there are several coves: Kapali, Abelike, and Porto Atheni (Fig. 10).

Porto Atheni is popular with yachtsmen. Two settlements have been preserved: Katomeri and Spartochori. Among the sightseeings of the island is the Papanikolis sea cave, where during World War II (1940–1941) the submarine "Papanikolis" was sheltered, and the Giovani cave, but they are accessible only from the sea.

(2) The strait separating the Meganisi Islands from the eastern coast of Lefkada Island. The southern entrance to the strait is clear, and in the north there are the islands of Skorpios, Sparti, etc. The width of the strait in the narrowest part is just over 900 m, and the depth is 45–110 m.

Messenian

Messenian (Kalamata) Gulf (*Messiniakos kolpos*) – the gulf in the extreme southwest of the Peloponnese peninsula of the Ionian Sea, Greece. It separates the Messinia Peninsula from the Mani Peninsula. The bay washes the southern shores of Messenia and the southwestern coast of the Mani peninsula in Laconia. The island of Venetico is situated at the western border of the bay, and in the southeast the bay is bordered by Cape Matapan. The shores of the bay are mountainous. The western coast is low and has fertile agricultural land, which is part of the plain that surrounds the northern part of the Pamisos river. The eastern shores are occupied by the southern foothills of the Taygetos mountain range, relatively rocky, inaccessible for climbing, in many places with steep cliffs going down sheer into the deep sea and numbering only a few settlements. Between the ridge and the Mesinia Peninsula, there is the Kalamata Plain. Many villages are situated at the places where mountain rivers and streams flow into the bay. The northern part is relatively densely populated with the largest cities of Kalamata and Messini, while the western part has many sandy beaches that attract tourists. On the northeastern shore of the bay, at its tip, there is the large port of Kalamata. The Pamisos River flows into the bay nearby the town. There are also many seaside resorts on the shores of the bay. On the west coast are the cities of Koroni, Longa, Chrani, on the east – Kardamyli, Stoupa, Agios Nikolaos. In 1986, near the city of Kalamata was the epicenter of a strong earthquake that wiped out many buildings from the face of the earth.

Messina

Messina – (1) one of three metropolitan cities in Sicily, Italy. The name is given for the largest city and dates back to the fifth century. BC when the ancient Greeks renamed the settlement of Zancle to Messena. In the ancient Roman period, the name "Messana" was used. It is located in the northeast of the island of Sicily – where it is separated from the mainland by the Strait of Messina. The coastline north of the strait is washed by the Tyrrhenian Sea, south of the strait – the Ionian Sea. The province also includes the Aeolian Islands. In the south, the province is bordered by the province of Catania, in the west with Palermo, in the southwest with the province of Etna. A significant part of Messina is occupied by the Peloritani and Nebrodi mountains. The area is 3247 km^2. The population is 650,200 people (2011). Messina is an important transport hub: the ferry crossing connects the island of Sicily and the mainland in the Strait of Messina. Connections with the Aeolian (Lipari) Islands are provided by regular flights from the port of Milazzo. Two railway lines run from Messina: one to the south – to Catania and Syracuse; the other to the west, to Palermo. In addition, the island of Sicily is supplied by electricity through a cable laid along the bottom of the Strait of Messina (Fig. 11).

(2) The administrative center of three metropolitan cities of the same name. The third largest city on the island of Sicily. The population is 231,700 people (2019). Messina is situated in the north-east of Sicily, on the western shore of the Strait of Messina, which separates the island from mainland Italy (Calabria region). The city stretches for 30 km along the Strait of Messina, and on land it is framed by the Peloritani

M

Fig. 11 Messina. (Source: https://www.goldenbaytour.com/UserFiles/Images/messina.jpg)

Mountains. In the fifth century BC, the city belonged to Athens and developed into a significant trading city. Around 493 BC it was renamed and named "Messene" in honor of the Greek Messene in the south of the Peloponnese. At the end of the fifth century BC famous tyrant Scythes ruled as a protege of the tyrant Gela Hippocrates. In 396 BC the city was ravaged by the Carthaginians, and rebuilt under Dionysius of Syracuse. In 312 BC Messene came under the rule of Agathocles, and in 288 BC was captured by the mutinous Campanian mercenaries in the service of Syracuse – the Mamertines ("sons of Mars"). From the time of the First Punic War, it remained in the hands of Rome. After the collapse of the Western Roman Empire, Messina was captured by the Goths, Byzantines, Saracens, and Normans. During the Crusades it was a port for the Crusaders and quickly flourished. Since the seventeenth century Messina began to decay. It was desolated and devastated by a terrible plague, earthquake, flood, revolutionary wars, and cholera. Messina also experienced the reign of the Angevins, getting rid of it during the "Sicilian Vespers." Having become Aragonese, the city shared the fate of the Spanish House until the eighteenth century. In the first 30 years of the eighteenth century Messina was ruled by the House of Savoy and the Austrian dynasty, and in 1734 their place was taken by the Bourbon dynasty, which ruled until the annexation of Sicily to the Kingdom of Italy (1860). The city was almost completely destroyed during the earthquake and tsunami on the morning of December 28, 1908. Then about 60,000–70,000 people died. Even the gravestones were damaged. Only two buildings remained intact: a prison and a psychiatric hospital. Fortunately, the ships of the Russian fleet were in the harbor, whose sailors saved thousands of Italian lives. The Baltic Fleet squadron wintered and conducted firing practice in Sicily, from where it came first to the port of Messina after receiving news of the disaster. In June 2012, a monument was inaugurated in honor of the Russian sailors who helped the townspeople who suffered from the 1908 earthquake. In 1909 the city was rebuilt, but suffered again in 1943 from American bombing, which led to the death of thousands of people. The main economic resource of Messina is the port – commercial and military – with several shipyards. Agriculture is developed. Lemons, oranges, tangerines, and other fruits and vegetables are grown. Wine is produced. The chemical, fruit-canning, textile industries, ship repair, and shipbuilding are well developed. Citrus fruits, vegetables, and canned fruits are exported. The University of Messina has been operating in Messina since 1548. Churches of the twelfth to thirteenth centuries and the cathedral of the twelfth to sixteenth centuries were preserved and partially rebuilt in the twentieth century; there are also fountains of the sixteenth century. The museum of marine fauna, the zoological museum of Cambria, the botanical garden, etc., are interesting.

3) A commercial port of Italy and a naval base (naval base) on the northeastern coast of the island of Sicily. It is situated in a natural, well-protected harbor of the Ionian Sea. Messina was founded by Greek colonists on the site of a settlement of the Sicels tribe around 730 BC and was originally called "Zancle" because of the shape of the natural harbor near it. The length of the mooring front is up to 3 km with a depth of 11 m. The entrance to the port of Messina harbor, 457 m wide, has a depth of 8 m. The naval base can provide basing of ships up to and including cruisers. The base is the headquarters of the naval command of the island of Sicily and is constantly deployed minesweeper forces. The total turnover of sea freight in the port of Messina is 0.15 million tons. There are five ferry berths in the southern part of the harbor. The port's ship-repair enterprises provide repair of ships and vessels of medium displacement.

Messina Earthquake *1908*

Messina Earthquake *1908* – occurred on December 28, 1908, in the Strait of Messina of the Ionian Sea between Sicily and the Apennine Peninsula. An earthquake with a magnitude of 7.5 is considered the strongest in the history of Europe. It began at about 5:20 AM at sea, at the bottom of the Strait of Messina. The tremors caused the displacement of the bottom areas. As a result, three tsunami waves with a height of up to 3 m battered the city of Messina at intervals of 15–20 min. In the city itself, three strong earth shocks occurred within 1 min. After the second shock, buildings began to collapse. As a result, the cities of Messina, Reggio Calabria, and Palmi were destroyed. In total, the earthquake affected more than 20 settlements in the coastal strip in Sicily and Calabria. The aftershocks continued in January 1909 (Fig. 12).

According to various estimates, a maximum of 200,000 people died. The most commonly cited figures are 70,000–100,000 people, including in Messina – 60,000, whose population was approximately 150,000 people. The first ships to arrive in Messina were the ships of the Midshipmen Detachment of the Baltic Fleet: the battleships "Tsesarevich," "Slava," the armored cruiser "Admiral Makarov," and then the armored belted protected cruiser "Bogatyr." Russian ships were part of a training detachment sailing in the Mediterranean Sea, and at the time of the earthquake were in the port of Augusta in Sicily. A few hours after the Russians the ships of the Mediterranean squadron of the Royal Navy of Great Britain arrived. The sailors took part in clearing the rubble, opened a camp kitchen, and transported the wounded to other cities on ships. Later, help came from many countries. In 1978, on the building of the municipality of the city of Messina, a memorial plaque was erected with the inscription "In memory of the generous assistance rendered by the crews of Russian warships ... to the inhabitants of Messina who suffered from the earthquake of December 28, 1908." On June 9, 2012, a monument to Russian sailors made of bronze was erected in Messina (sculptor – Pietro Kufferle, foundry – Carlo Nocherdi), created back in 1911, which depicts sailors rescuing people from under the ruins, with an inscription on the pedestal "To the August sailor, the representative of the heroes of mercy from the grateful sons of their homeland." (Fig. 13).

In the Central Naval Museum of St. Petersburg, there is a certificate of gratitude that reads "To you, the generous sons of a noble land, whose heroism will go down in history, who were the first to come to the aid of those many who were threatened with certain death from the fury of the earthly firmament ...." The order to help the Sicilians was given by Nicholas II, who, in particular, ordered to provide the Italians with any help. Actually, for this a bust in the city of Taormina was erected in his

Fig. 12 Messina earthquake. (Source: https://starsity.ru/wp-content/uploads/2019/12/zemlyatresenie-v-messine-1908.jpg)

M

Fig. 13 A Monument to Russian Sailors. (Source: https://f.otzyv.ru/f/11/09/88944/090218194859 88459731.jpg)

honor. In memory of the rescuers in Messina, the following streets are named: the street of the Russian heroes-sailors of 1908, the street of the Russian sailors, the street of the Russian sailors of the Baltic squadron. In 1910, the Italian government awarded all Russian sailors with small silver medals with the inscription "In memory of the calamity that befell Messina and Calabria" for rescuing people. The Russian fleet was awarded with a large gold medal, and the ships participating in the rescue with large silver medals. The commanders and physicians of the ships were especially honored.

Messolonghi

Messolonghi (Missolonghi) (*Messolongion*) – (1) city and port of the same name at the entrance to the Gulf of Patras of the Ionian Sea, Greece. The administrative center of the Aetolia-Acarnania nomos. The population is 12,700 people (2011). The city is located on three islands at the mouth of the Achelous and Evinos rivers on the low marshy northeastern shore of the Missolonghi Lagoon. It is known as one of the centers of the Greek liberation uprising against the Turkish yoke in 1821–1829, and is also associated with the name of an outstanding poet, national Greek hero, Lord Byron, who died there. The city begins with the Gate of Exodus – it was through them that on April 12, 1826, 9000 men, women, and children tried to break out of the Ottoman blockade, in which they had been for more than a year. Near the city near Mount Zygos, an ambush of ruthless Albanian mercenaries awaited them. As a result, only less than 2000 were saved. Near the gate is the "Garden of Heroes" – Kipos ton Iroon, where the remains of the city's nameless defenders rest. This status was enshrined in law in 1937, making it the only hero city – the "Holy City" in Greece. The historic exodus of the "Free Besieged" became the personification of the free and independent spirit of the Greeks, making the city a sacred symbol of this spirit (Fig. 14).

Here in 1881 a statue of Byron was erected, under which the poet's heart was buried (according to a number of sources, his lungs were buried). His other remains were transferred to his family estate in England. Byron arrived in Mesolongion in January 1824. He was proclaimed commander-in-chief of the 5000 garrison. He drilled soldiers, and trained in shooting and rowing in the marshy lagoon that surrounded the city,

Fig. 14 Messolonghi. (Source: https://www.greece-travel-secrets.com/images/Missolonghi-Gulf-of-Corinth-Greece-2.jpg)

where he caught a fever and died on April 19, 1824. The house where Byron lived and died was destroyed in World War II. The city has a small museum dedicated to the Greek Revolution. A small "Byroniana" – scattered objects – and the original "Hymn to Freedom" by Solomos – the words of this work have become the Greek national anthem, – are presented here. At the foot of Arakynthos there is the Monastery of Saint Simeon (1740), where the surviving defenders of the city found refuge after the Exodus of 1826. In the northwest of Megalonisi lie the ruins of the amazing city of Aetolia – Pleuron (Plevrona). Not far from Stamna there is the Chapel of Saint Agatha.

(2) Port, accessible for vessels with draft up to 3 m. Depth in the port is 7 m. It is organized in the northeastern part of Missolonghi Lagoon. Through it, trade is carried out between the settlements of the northern coast of the Gulf of Patras and other ports in Greece. The port has embankments with a total length of 600 m. A canal with a length of approximately 5 km, its width is 70 m, and the smallest depth is 5 m. On the eastern side of the entrance to the channel lies a low island Tourlida. It is connected to the city of Mesolongion by a 4.5 km long bridge, under which only boats can pass.

(3) A lagoon, which is a large shallow water with numerous islets and muddy banks, stretching along the northern coast of the Gulf of Patras for more than 20 km and almost the same distance inland. The lake is separated from the bay by a chain of low sandy islets and reefs, the largest of which are Tholi and Agios Sostis. The lake is rich in fish. The northern shore of the lake is low and marshy. The city of Messolonghion is located on the northeastern coast. The northern part of the lake is connected by a narrow strait with Aitoliko Lagoon. The lake is accessible for navigation only with knowledge of local conditions. In the north-eastern part of the lake, there is a port, through which trade is carried out between the settlements of the northern coast of the Gulf of Patras and other ports of Greece. Depths in the port are 5.4–8 m.

Metaponto

Metaponto – (1) a city located in the western part of the Gulf of Taranto of the Ionian Sea in the region of Basilicata, Italy. The population is about 1000 people. Founded by The Achaeans between the eighth and seventh centuries BC. When Pythagoras was expelled from his native Crotone, he settled in Metapontum. The dusk of Metapontum occurred during the Second Punic War. This was Hannibal's headquarters. The Phoenician commander drove all the inhabitants of Metapontum out of the city, taking revenge on the Romans, who in 207 BC captured Tarentum and sold its inhabitants into slavery for supporting Hannibal. Among the sightseeings of Metaponto are the house of Pythagoras and the ancient Heraion – the temple of Hera of Metaponto, where in the Middle Ages, before the next Crusade, knights-paladins gathered. Near Metaponto there is a beautiful sandy beach – Metaponto Lido.

(2) The plain which is situated on the Ionian coast of the Gulf of Taranto, Basilicata region, Italy. The plain stretches for 35 km from the city of Metaponto to Nova Siri, and in depth – to the hills of Matera. The plain is often referred to as "Southern California" and is considered the pride of the area's economy. From here, fruits and vegetables are exported to Europe and all over Italy, since they are ripen here 2–3 months earlier than on the plains located to the north.

Methoni

Methoni – small town on the southwestern tip of the Peloponnese peninsula, Messenia region, Greece. An outstanding natural harbor. The population is 1100 people (2011). The main attraction is the imposing Venetian fortress, built on the place of ancient cities in the Mycenaean era. In 1209, Methoni was known as the "trench of Venice." The Methoni fortress is built on a bay bar that shoots out into the sea. It is separated from the mainland by an artificial moat. On a small islet of Bourtzi, connected to the fortress by a bridge, the Turks built the eight-sided Bourtzi tower to strengthen the vanguard of the fortress, as a watchpost and a lighthouse for ships. Here you can see the remains of a Turkish bath, a church, and in the middle of the fortress there are fortifications. In ancient times, Methoni was known as Pedasus. Homer lists it among the seven cities that Agamemnon offered to Achilles. Pausanias and Strabo so far call it Methoni. During the Roman period, the city gained independence from Emperor Trajan and was strengthened by the construction of fortifications. By the late Christian period, Methoni was known as a trading center and a base for the provision of food and water for ships. In the eleventh century Methoni was in the hands of the Byzantines, and then in 1204 it passed into the possession of the Franks, and in 1209 Methoni passed again under the control of the Byzantines. During this period, Methoni became a major trade center, as it was an extortion stop for all Venetian ships going to the East. In 1500, after decades of siege, the city was occupied by the Ottomans, led by Sultan Bayazet. The first period of the Turkish rule lasted until 1696, when the city returned to the possession of the Venetians. However, in 1715 the Ottomans took possession of Methoni again. During the Greek Revolution, the rebels were unable to capture Methoni. In 1825, the Egyptians, led by Ibrahim Pasha, landed at the port and Methoni became a bridgehead during their campaign in the Peloponnese. In 1829, the French, who liberated the city as well as the entire Peloponnese, settled in the fortress. At the same time, the settlement was moved outside the fortress. An arched stone bridge connects the entrance gate of the fortress with the city – this is the creation of the French general Maison, who planned the development of the cities of Methoni and Pylos. The southern battlement ends at the Sea Gate leading to the harbor.

Milanio Naval Museum

Milanio Naval Museum (Museum of Navigation) – it is situated in the resort village of Tsilivi on the island of Zakynthos. Numerous items related to the sea are presented. Along the way, in chronological order, covering a 4000-year period, the history of the navigation of Hellas, Greece is revealed.

Milo of Croton

Milo of Croton (*Milon Crotone*) – the famous athlete and wrestler from Croton of the

Antiquity era, who won numerous victories. In 532–516 BC he won 6 Olympic Games and 25 other sports events. There are many legends about his unusual strength: he could lift a bull on his shoulders.

Molos Gulf

Molos Gulf (Kolpos Molou) – a bay that penetrates deep into the island with a blunt wedge 5.5 km southwest into the eastern coast of Ithaca island, Western Greece. It cuts the island nearly into two parts: north and south. The southern coast of the bay is low. The northwestern coast of the bay is high and almost straight. It is the eastern slope of a mountain range that runs along the isthmus connecting both parts of the island. The width of the isthmus is 750 m. The bays of Skinos and Vathy penetrate deep into the southeastern coast of the bay. The depths in the central part of the bay are more than 100 m, in largest part of it – 50 m. At the top of the Vathy bay there is the administrative center of the island – the port city of Vathy.

Montante

Montante – the local name for the constant current in the Strait of Messina, directed to the north. When it flows along the shores of the strait, the countercurrent flow of the "Bastardi" is observed. The Montante current, passing along the coast of the island of Sicily, heads towards the coast of the Calabria peninsula to the area located south of the harbor of Reggio Calabria. Further, it goes along this coast to the north. The further to the north, the higher is speed, reaching 4–5 knots on the transverse member of the harbor of Reggio Calabria, and 6–7 knots in summer. In places where the Montante current meets countercurrents, especially in the presence of uneven depths, maelstroms are formed, which have the local name "Garofali." The strongest ones are the "Scylla" maelstrom off the mainland coast of Italy and the "Charybdis" maelstrom near the village of Torre Faro.

Morea

Morea – the medieval name of the Peloponnese peninsula. During these times a resemblance to a mulberry leaf (mulberry) was guessable in its outlines.

Mount Etna

Mount Etna (*Etna*) – volcano. "The pillar of heaven, Etna covered in snow" – this is how the great ancient Greek poet Pindar called it. One of the most active volcanoes in the world. It is situated in the northeast of the island of Sicily, near the coast of the Ionian Sea, Italy. Its height is 3343 m, and it is the highest volcano in Europe. The forming of Etna began about 600,000 years ago. About 500 thsd. Years ago the volcano acted as an underwater one, but about 150,000 years ago it rose to the surface. For 8000 years BC parts of the summit cone exploded and collapsed into the Ionian Sea, causing a great catastrophe. Waves up to 13–15 m high struck the coast of western Greece, up to 8–13 m – on the shores of Egypt and present-day Libya. After another 6000 years, the explosion of the cone was repeated, which forced the local tribes to move to the west of the island. The first historically recorded eruption of Etna occurred in 693 BC, which for the first time destroyed the city of Catania, founded three and a half decades earlier. Since then, the volcano's lava has repeatedly approached the walls of the largest city on the eastern coast of the Ionian Sea of Sicily. Thus, during the eruption, 1000 people died in 1536, in 1669 20,000 people and in 1980 4800 people (Fig. 15).

Eruptions are often accompanied by an earthquake, as happened in 862 and 1169 – then the number of victims exceeded 10,000 people. The eruption of Etna can last for years, as it did in 1614, when it lasted 10 years.

Fig. 15 Mount Etna. (Source: https://ychef.files.bbci.co.uk/1376x774/p0885fws.jpg)

Fig. 16 Mourtos Island. (Source: https://www.fodors.com/wp-content/uploads/2018/06/Greece-Best-Beaches-Myrtos-Cephallonia-Lonian-Islands.jpg)

Mourtos

Mourtos (Syvota) – an island located in the southern part of the Corfu Strait near the Megalo Mourtemeno Peninsula, Greece. The island is rocky, 118 m high. It stretches for 2 km to the northwest, and is approx. 500 m. Covered with dense forest. Its southeastern part is connected to the island of Agios Nikolaos by a sandbank with a depth of 3.6 m above it and to the mainland. There is a small village and fishing port (Fig. 16).

Murro di Porco

Murro di Porco (Capo Murro-di Porco) – a cape at the southeastern tip of the Maddalena Peninsula, Sicily, south of the east coast city of

Fig. 17 Mussels. (Source: https://guildcorp.ru/wp-content/uploads/2021/06/cdd73e77f92bda688ee99a5edef31649.jpe)

Syracuse. The peninsula itself is composed of steep cliffs of 15 m high. The eponymous lighthouse in the form of a red tower with a house was placed on the cape.

Museo del Mare

Museo del Mare – is located in the city of Messina, Italy. The basis of the exposition is made up of old navigation (swimming) equipment and tools of labor of fishermen and sailors.

Mussels, Mytilus

Mussels, Mytilus – Mollusks, family of marine bivalve mollusks. The shell of mussels is wedge oval, smooth. Length is up to 20 cm. Coloring is attached to a solid substrate, as well as to other mussels, forming their intergrowths. In the mantle cavity of mussels, water is circulated, which is necessary for respiration and the entry of food particles. Large mussels can leak out through the mantle cavity up to 70 l per day. Mussels purify water from suspended particles, which are then deposited on the ground. Sometimes mussels are called "oysters for the poor." Mussels meat is nutrient rich, and eaten in boiled and canned form. Some types of mussels are bred artificially. It is widely distributed in temperate and subtropical waters of the Northern Hemisphere. Edible mussels sometimes form solid settlements – mussel banks. It is a component of fouling of underwater parts of port facilities and vessels yellowish-green. With the help of byssus mussels (Fig. 17).

N

National Marine Park of Zakynthos

National Marine Park of Zakynthos (National Sea Park – Reserve of Greece) – located in the Gulf of Keri on the southern coast of Zakynthos. Founded in 1999, it is the first marine park in the Mediterranean to protect sea turtles. It occupies part of the beach and the bay of Laganas, as well as two small islets south of Zakynthos – Marathonissi and Peluso. The area is 135 km^2. It is the part of the Natura 2000 project. The park represents a typical Mediterranean ecosystem: sand dunes and beaches washed by the clear waters of the Ionian Sea and bordered by lush vegetation and pine groves. It is home to rare relict turtles, dolphins, seabirds, porcupines, and iguanas. On the coast of the island you can see huge sea turtles *Caretta caretta* and a rare Mediterranean marine seals species – monk seal. Fishing activities are strictly prohibited to protect the ecosystem.

Naupaktos

Naupaktos (Nafpaktos, Nafpakt) *(former Lepanto in medieval times)* – the third largest city in the prefecture Aetolia-Akarnania. It is located on the north shore of the Gulf of Corinth of the Ionian Sea, 15 km north – east of Patras, Greece. It is a commercial resort town with a Venetian castle, one of the most beautiful in Greece. The city has two excellent beaches – Gribovo and Psani, surrounded by groves that descend from the hills to the water. Sandy beach Gribovo, which was awarded by the EU Blue Flag, is located in the eastern part of the city and is famous for its purest water (Fig. 1).

For many centuries the city has been the political, economic, and military center of Greece. Two picturesque ancient towers guard the entrance to the old Venetian port – the smallest and most beautiful port in the Mediterranean. On the western side of the port there are two monuments to the defender of the city Jorgos Anemojanis, who was martyred during the siege of 1821, and a little further, to Miguel Cervantes, who not only defended the city against the Turks but also lost his left arm in these battles. This, fortunately, did not prevent him from writing his famous "Don Quixote" later. The port is located in a semicircle at the foot of the hill, on which the city stretches like an amphitheater. At the top of the hill, a fortress (castro) rises like a majestic crown, which is one of the best preserved fortresses in the country, as well as the most beautiful of them. It was built by the Venetians in 1407 when they captured the city. "Lepanto Guard", as the fortress was called before, impresses with its complex architecture, imposing walls, and five tiers. Legend says that somewhere in the inner part of the fortress there is a secret tunnel, which gave the inhabitants of the city a safe exit to the harbor in difficult times. Below are the ruins of Vezir Tzami (the vizier's mosque) and the house of the leader Kitsos Tzavellas. The heroic and romantic atmosphere has been preserved here, which has

I. S. Zonn, *The Ionian Sea Encyclopedia*, Encyclopedia of Seas,
https://doi.org/10.1007/978-3-031-08206-1

Fig. 1 The city of Naupaktos. (Source: https://www.nafpaktos.gr/sites/default/files/fin_slid_1_1.jpg)

been inherent in it since ancient times. This is a matter of special pride and concern for its residents. In 1571, the name of the city became famous in all parts of the world, when, during the famous Battle of Lepanto, the united European, Christian forces defeated the Ottoman fleet. The nearby built Rio-Antirrio bridge (q.v.) again brought the city to the forefront. The Venetian castles of Rio and Antirrio, which once guarded the entrance to the Gulf of Corinth, have survived. They were built back in 1686 during the campaign of the Venetians to Greece. The city has an interesting private collection of relics from 1821 in the museum at Agnostu Stratioti (Unknown Soldier) Square, the Archaeological Museum near the Old Port. The Holy Monastery of St. John the Baptist Vomvokou, which dates from 1695, the Monasteries of Panagia Ambelakiotissa, and the Holy Transfiguration of our Lord and Savior Jesus Christ are located in the mountains around Nafpaktos.

Nautical Museum

Nautical Museum (Museum of Nature and Navigation) – a recently created museum in the city of Fiskardo on the northeastern tip of Kefalonia. The volunteers who work here study the environment and the underwater world of the Ionian Sea.

Navagio

Navagio – one of the bays of the island of Zakynthos in the Ionian Sea, Greece (Greek. Shipwreck Bay), which is rightfully considered the most beautiful and unusual place in the world, the visiting card of the island of Byzantium. The rusting wreck of an English cargo ship "Panayotis", built in 1939, lies in the bay with dazzling white sand. She was thrown ashore during a violent storm in 1980, which gave the bay its name.

Navarino

Navarino – a bay of the Ionian Sea, one of the most comfortable, sheltered from the wind and waves in the Mediterranean. It penetrates deep into the western coast of the Peloponnese, Greece.

The bay looks like a semicircle and is slightly open. From the west of Sphacteria island there are two entrances to it: the Northern one, through the Sikia Strait, and the Southern one, which is the main one. The eastern coast of the bay is rocky; it gradually rises inland. The northern coast is sandy, with a vast Dalian lagoon and bogs lying close to the coastline. From the west, the bay is bounded by the high and rocky coast of Sphacteria Island. Several rivers flow into the bay, flowing through the valleys between the hills. The largest of them is the Gialova River. The depth in the middle part of the bay is 40–50 m. The depth in the southern passage in the middle is 55–70 m. It is known as the place of the glorious victory of the Russian fleet in 1827, when the Russian squadron helped the Greek people who fought to overthrow the Turkish yoke.

Naxos

Naxos – the first Greek colony on the northeastern coast of the island of Sicily south of Taormina, founded in 734 BC, at the foot of Mount Etna. The settlers came from Chalkis and Euboea, as well as from the island of Naxos in the Aegean Sea. In 415 BC, during the naval expedition of Athens to Sicily, Naxos sided with Athens. In 403 BC, it was destroyed by Dionysius I. The Greek column – the oldest on the island of Sicily – the remains of the walls, and the temple of Aphrodite can be specially noted among its sightseeings.

Neda

Neda – river in Greece, in the Peloponnese. It has its source at the village of the same name on Mount Lykaion. It flows on the border of Messinia and Elis, of the Ionian Sea. The length of the river is 32 km. Named for the nymph Neda, who raised Zeus. According to legend, it was created by Rhea after the birth of Zeus to wash the baby. In the village of Figaleia there is a waterfall 50 m high. By the river there were the ancient city of Phigalia, the ancient city of Lepreon, and the Temple of Apollo in Bassae. As we go down, away from the village of Platania, there is a stone bridge relating to the period of Turkish rule.

Norway Lobster, *Nephrops norvegicus*

Norway Lobster, *Nephrops norvegicus* – species of lobsters from the order of decapod crustaceans. The main specificities of the Norway lobster are two oblong claws. Females reach sizes up to 20 cm; males reach up to 24 cm in length. Most of those caught today only occasionally exceed 15 cm. It dwells on the soft areas of the seabed, at a depth of 40–250 m, where it digs its own caves and passages. It leaves its refuge only at night to search for food. Every second year from March to November, females lay up to 4000 eggs, which are fertilized by the male only after they are stuck on the tail of the female. The female carries these eggs with her for up to 9 months. It is considered a delicacy, while its thin claws, unlike other types of lobster, do not contain meat suitable for eating. Norway lobsters are often offered under various names. In France, it is called "Langoustine," which brings it closer to much more expensive langoustes. In Germany and Austria, Norway lobsters are also often served as a "langouste tail".

Noto Gulf

Noto Gulf (Golfo di Noto) – one of the largest bays of the Ionian Sea on the east coast of the island of Sicily, Italy. It is located in the southeast of the island and is bounded in the north by Cape Murro di Porco and in the south by Cape Portopalo.

Nydri

Nydri (Nidri) – famous historical and popular resort town on Lefkada island, Greece. It is located on its eastern bank at the entrance to the

Fig. 2 Nydri (Nidri). (Source: http://turistua.com/pictures/gallery/36466.jpg)

Vlycho Bay, which penetrates deep into the shore. It is separated from the Ionian Sea by the Yeni Peninsula. There is a constant ferry service with the Meganisi island. From here you can also go to many excursions to nearby small islands which can be seen from the embankment – Sparti, Skorpios, and Maduri. The famous German archaeologist W. Dörpfeld believed that Odysseus reigned not in Ithaca, but in Nydri, building on the results of his excavations and the discovered burial of the Bronze Age. The perseverance with which he defended his theory earned him the gratitude of local residents, who erected a monument to him on the embankment (Fig. 2).

O

Odysseus

Odysseus (лат. Odysseus, Ulixes) – in Greek mythology, the brave hero of the Trojan cycle of myths, the king of Ithaca, the son of Laërtes and Anticlea. In post-Homeric mythology, he was also considered the son of Sisyphus. Odysseus prompted all the suitors of the beautiful Helen (daughter of Zeus and Leda), to whom he wooed, to swear that in case of danger they would come to the aid of her chosen one. When, after the abduction of Elena, Odysseus was supposed to take part in the war against Troy, he pretended to be insane, since the Oracle predicted to him that he would return only 20 years later (Fig. 1).

Palamedes, the son of King Nauplius, exposed Odysseus's alleged madness. Then he, leaving his wife Penelope, the son of Telemachus and his property, went to war, where he bravely fought against Troy, helping the Greeks in difficult situations with his wisdom and eloquence. By cunning he achieved the death of Palamedes. Palladium (a statue of the goddess Athena) was kidnapped from Troy. Gave advice to build a wooden (Trojan) horse. After the conquest of Troy, during his 10-year return home, Odysseus experienced many adventures with the Cicones (Thracian people), lotus-eaters (a peaceful tribe that fed on the lotus), and the Cyclops Polyphemus. Having almost reached Ithaca, Odysseus's companions untied the bag of opposite winds that Odysseus received from the lord of the winds Aeolus, and the ships were carried back to the sea. After only one ship of Odysseus escaped from the cannibals of the Laestrygonians, he descended into the realm of the dead and asked about the future the shadow of Tiresias (the legendary blind soothsayer from Thebes). After that, he preserved the companions from the mind dulling singing of sirens and led the ship through the strait, guarded by Scylla and Charybdis. The companions of Odysseus, despite the warning of Tiresias, killed the solar bulls of Helios on the island of Thrinacia, thus the ship sank and Odysseus alone escaped to the island of Ogygia, where he spent 7 years with the nymph Calypso (daughter of Atlas). On the way home, he was rescued during a shipwreck by the veil of Leucothea (sea deity). Having reached the Isle of the Phaeacians, he was brought by Nausicaa to her father, the king of the Phaeacians, Alcinous, and with his help he finally reached Ithaca. Subsequently, Odysseus was killed by accident by his son Telegonus, who did not recognize his father. Since antiquity the adventures of Odysseus have been a favorite subject of literature and art (cinematographers, artists, composers). In the city of Stavros on the island of Ithaca, a statue of Odysseus is settled.

I. S. Zonn, *The Ionian Sea Encyclopedia*, Encyclopedia of Seas,
https://doi.org/10.1007/978-3-031-08206-1

AU2

Fig. 1 Odysseus. (Source: https://www.historytoday.com/sites/default/files/2021-06/Ulysses.jpg)

"Odyssey"

"Odyssey" – an epic poem created by Homer, describing Odysseus' wanderings and his return to his homeland.

Olive

Olive (*Olea europaea*) – evergreen, subtropical tree. It has been cultivated for olive oil since ancient times. Homeland – the southeastern Mediterranean. The cultivated form of the European Olive is grown in all Mediterranean countries, on the Black Sea coast of Russia, in Georgia, Azerbaijan, Turkmenistan, Iraq, Iran, Pakistan, and northern India. In 1560 it was brought to America, where it is cultivated mainly in Peru and Mexico. It was first cultivated in Greece, where it is grown in vast numbers up to now. Olive is considered one of the longest-livers of the plant life. Average life expectancy is 300 years, maximum 2000! The growth of plant is slow. The tree has a height of 4–5 m, and an evergreen shrub – 1–3 m. Olive is a perennial plant due to its ability to regenerate itself. It is a drought-resistant horticultural species that can withstand short-term frosts down to −10°C. Olive trees grow well in soils with sufficient lime content, on rocky sandy soil. Their roots are so strong that trees can even grow in rocks. The root system is highly branched and by several times exceeds the projection of the crown of the tree, although it is located at an average depth of 80 cm. This allows Olive to withstand droughts. The root system and trunk form a thickening – a cervix, which very often enlarges, grows. Young shoots grow from it every year, which must be constantly removed so that the tree does not turn into a huge bush. The height of olive's trunk does not usually exceed one meter. Covered with ash-gray bark, gnarled, twisted, and in old age, it is usually covered by holes. The branches are knotted, long, inclined in some cultivars. The leaves are simple, leathery, oblong, green above, silvery-gray below (Fig. 2).

At the edges, the leaf is slightly incurved (therefore, the area of heating by the sun's rays decreases). Such a specific leaf structure helps the plant to withstand severe droughts. During wet mild winters, Olive accumulates a large amount of reserve starch in the leaves, which serves it as a good help in spring and dry, waterless summers. The evergreen leaves on the Olive tree are gradually replaced, with new leaves beginning to form during the period from spring to autumn. Leaf lifespan is 1 year, very rarely reaches two. Leaves do not fall off in winter. A tree growth bud at each leaf base can be dormant for a long time. It grows only if necessary – with strong pruning, damage from hail, etc. It is a very valuable property of the

Fig. 2 Olive. (Source: https://justfruitsandexotics.com/wp-content/uploads/pkytnrj82c4fxqroalgo-scaled.jpg)

olive tree to recover every time. On many islands in Greece there are olive groves with preserved trees that are 1400–2000 years old. The trunks of the trees are gnarled, with voids. In winter, these voids are filled with water, and in summer they use its moisture. Previously, oil was a strategic raw material and was stored in large earthen vessels with a volume of 400 liters (you can see them in museums). Now almost every islander has his own olive garden. From a young tree, you can collect olives for 3 liters of oil per season. Depending on the climate, olive trees bloom from late April to early July. Fragrant flowers are very small – 2–4 mm long, whitish, with two stamens, located in the leaf axils in the form of paniculate brushes. One inflorescence contains 10–40 flowers. Olive fruit is a pyrenocarp, most often elongated-oval in shape, 0.7–4 cm long and 1 to 2 cm in diameter, with a pointed or obtuse spout, with a fleshy pericarp containing oil. The color of the fruit pulp differs depending on the type of tree: it can be either green, or black, or deep purple, often with an intense waxy bloom. The nucleus is very dense with a grooved surface. Fruit ripening occurs 4–5 months after blossoming. Olive tree become productive only after 20 years. The tree bears fruit once every 2 years. The fruits are harvested from September to December. The 90% of Olives are used to make olive oil, which, even unpreserved, has a long shelf life, which is of great importance for the Mediterranean. Other olives are pickled with or without nucleus. Since ancient times, people have eaten the fruits of the olive and made olive oil of them. Olives are rich in fats. The oil yield on absolutely dry matter is 50–80%. Fruits are rich in proteins, pectins, sugars, vitamins: B, C, E, P – active catechins, and contain salts of potassium, phosphorus, iron, and other components. In addition, they contain carbohydrates, phenol carboxylic acids, pectin substances, and triterpene saponins. Olive oil is the main product for which this crop is mainly cultivated. However, the fruits of the Olive are widely used in the canning industry for the manufacture of canned food of green fruits, and of black, dry-salted olives. Provencal olive oil is used in the food industry for the production of delicious fish products (sprats, sardines). Canned olives, dry-salted black olives, and especially stuffed olives have a piquant taste, and they are an appetizer, delicious canned food, complementing the range of food products, first of all, having a healing value. The wood is greenish yellow, heavy, strong, and wave growth timber, which lends itself well to polishing and is used for making furniture. It is also appreciated by woodcarvers, used for inlays and the creating of expensive turnery and woodwork. There are attempts to replace quin by an olive bark. Olive leaves infusions normalize blood pressure and respiration. The raw pulp of the fruit contains up to 80% of non-drying oil, which includes unsaturated fatty acids – oleic (75%), linoleic (13%), and linolenic (0.55%). Unlike animal fats, they are not harmful, but also healthful to the body. The fact is that the contained acids in them – the main component of vitamin F – are necessary as a building material for cell membranes, and the body itself synthesizes them only partially. Olive oil is used in the cosmetic industry. The second and lower grades of oil are known under the name of wood oil and are used for lubricating machines and in soap making. Plants are a good ameliorant on terraced mountain slopes steeper than 10–12°. They make the soil resistant to erosion and subsidence, which is very important in stopping landslides, soil erosion, and wasteful water discharge from precipitation. The olive branch is a symbol of peace and truce. According to the Bible, it was brought to Noah by a dove as a sign that the anger of God against people has ceased and the Flood has stopped. The olive tree was the first to grow after the flood, and therefore its branch became the message of peace between God and man. In Islamic tradition, the olive is the tree of life, one of the two forbidden trees in paradise. The dove carrying the olive branch is depicted in the emblem of the World Peace Congress. The olive branch has found wide application in heraldry, where it has a similar meaning.

Onassis Aristotle Socrates (1906–1975)

Onassis Aristotle Socrates (*Onassis Aristotelis Socrates*) (1906–1975) – Greek ship owner, billionaire. A shipping magnate and international businessman who has built a fleet of supertankers and cargo ships larger than the fleets of many countries. Born in Smyrna (now Izmir, Turkey) into a fairly wealthy family of tobacco dealers. In 1922 it lost almost all of its fortune when Smyrna, a Greek city at that time, was captured by the Turks. Having fled from Smyrna to Greece, the family sent the Onassis to Argentina in search of a better life. He had $ 63 in his pocket. In Buenos Aires, Onassis became a night port dispatcher and, with the help of family friends, started a daytime tobacco import business. His tobacco shop, "Oriental Tobacco Import," was on the main street of Buenos Aires. He increased the use of imported oriental tobacco in Argentina from 10% to 35% and 2 years later earned $ 100,000 from a 5% commission from the tobacco trade. His business activities expanded due to cigarettes and other consumer goods production. At the age of 25, Onassis made his first million dollars. Having become rich, Onassis bought the seat of the Greek consul in Buenos Aires, and entered into a trade agreement with Argentina in 1928. In 1932, at the auction, where six dry cargo ships of the bankrupt Canadian company were put up as a lot, there were no buyers, and Onassis offered his amount in 100 times smaller than the initial one and bought them. Onassis entered the shipbuilding business, buying his first six ships for $ 120,000. He used two of these ships until ships with a higher carrying capacity appeared, and then bought others. Onassis built his first tanker in 1938 and acquired two even larger ones by the start of World War II. In society, Onassis was treated with caution: everything around him was saturated with power and strength. He could smile, but his gaze remained cold. Friends often called Aristotle a monster: he did not have any moral principles – he supplied the warring countries with oil, and during World War II he did business with the Nazis. He was cruel to women too. His empire grew and his fleet expanded throughout the 1940s and 1950s. In the mid-1950s he bought 17 new tankers in 1 year. In 1953, he also paid $ one million for a control stock, acquiring almost all the entertainment establishments and the "Society of Sea Bathing of Monaco" (Société des bains de mer de Monaco (SBM)). Since 1957 to 1974 he has been the owner and operator of Olympic Airways, the Greek national airline concessioned to him by the Greek government. His first wife in 1946 was Athena Livanos, the daughter of the ship magnate Stavros Livanos. In marriage, they had two children: son Alexander and daughter Christina. This marriage ended in divorce in 1960. For a long time, Onassis was close to the Greek opera diva Maria Callas (he did not allow her to leave the child). In 1968 he married Jacqueline Bouvier Kennedy, widow of the 35th US President J. Kennedy. It was not a marriage of love, but of convenience. Onassis' only son, Alexander, died on January 21, 1973, in a plane crash at the age of 25. The luxury yacht named after his daughter "Christina O," heiress of 2/3 of his estate, was called "floating palace" by some people and served for several years as a permanent residence. It was decorated with mosaics and marble, golden taps in the bathrooms, a swimming pool that was transformed into a dance floor. Christina herself was married four times (including marriage to Sergei Kauzov, a citizen of the USSR and an employee of "Sovfrakht." The marriage was concluded in the summer of 1978 in Moscow and lasted for 16 months). She died on November 19, 1988, in Buenos Aires at the age of 37 under mysterious circumstances. After Christina's death, the empire of Onassis passed to her

Fig. 3 Onassis Aristotle Socrates. (Source: https://drunkguest.com/wp-content/uploads/2018/12/aristotelonasis2.jpg)

daughter Athina Roussel (born in the last, fourth marriage of Christina Onassis, concluded in 1983 with the French Thierry Roussel). According to the European Association of Myasthenia gravis, Onassis is one of the most famous people in the world with such a disease. This fact is very important for a huge number of patients with this disease all over the world, since it demonstrates a person's capabilities, even with such a serious neurological disease, to lead an active lifestyle, engage in effective business activities, and have a rich social life. According to the will of Onassis, he left 55% of the inheritance to his daughter Christina Onassis, and transferred 45% to the establishment of the Alexander Onassis Foundation (Fig. 3).

Operation "Marita"

Operation Marita ("Marita") – a cryptonym for the plan for the capture of Yugoslavia and Greece by the troops of the Axis countries during the World War II.

"Orange coast"

"Orange coast" – is a coast of the Ionian Sea in the area of Calabria, Italy. The center of this coast is the city of Soverato. It was named "Orange coast" due to numerous plantations of orange trees stretching to the north and south from the sea shore. The traditional handicraft masterpieces created in the area of "Orange coast" include works on wrought iron in Cropani and Borgia, earthenware from the towns of Tiriolo and Cerva, Cropani embroidery, leathers from Cellia and musical instruments from Tiriolo.

Ortega

Ortega (Ortigia) – an island (also known as the Island of Partridge; the name of the island originates from the ancient Greek ortyx (ὄρτυξ), which means "Quail") is situated in close proximity to the mainland of the city of Syracuse, Italy, and connected to it by a bridge built in 1870. Its surface is completely occupied by the city buildings. It separates the two harbors of the port of Syracuse: Porto Mamoreo and Grand Harbour, Syracuse, connected by the shallow Darsena Channel it was here, where Greek colonists landed and established settlements in 734,733 BC. The depths at the entrance to the harbor are 20 m. Very few buildings preserved on the island in the oldest part of the city. Among the surviving objects there are: the Doric Temple of Apollo (sixth century BC), the oldest in Sicily, discovered after excavations in 1938–1943; the Duomo Cathedral is one of the only ones in Italy that combines a pagan sanctuary and a Christian temple. The cathedral was damaged during an earthquake in 1693, but was rebuilt. The main chapel of the cathedral contains relics associated with St. Lucia – a huge statue 4 m high, weighing 1 ton, depicting a Syracuse martyr and several bones from a foot; the Arethusa fountain is a natural reservoir located a few meters from the Ionian Sea, formed by the confluence of two small rivers. Papyrus grows here – the herb that once covered most of Sicily; there is a papyrus museum in Syracuse; Castello Maniace – a medieval palace – a fortress, built in 1239; Roman amphitheater of the third and fourth centuries. On the southern tip of the island, the Lighthouse Castello Maniace is installed (Fig. 4).

Fig. 4 Ortega. (Source: https://static.dezeen.com/uploads/2017/06/made-labs-design-workshop-sicily-competition_dezeen_2364_col_14.jpg)

Othoni

Othoni (*Othonoi, Fanó (um.)*) – the largest of the three large islands of the Diapontia Islands in the Ionian Sea, located 8 km northwest of the island of Kerkyra, Corfu, 10 km northwest of Mathraki, and 15 km west of Ereikoussa. The area of the island is 10.4 km^2, and the population is about 392 people (2011). The island is 5.6 km long, with a maximum width of 3.6 km. The length of the coast line is 17 km. The island is covered with dense vegetation. It is mountainous. The highest point is Mount Imerovigli (also Merovigli) (393 m). In the west, coastal cliffs of 500 meters drop into the sea; in the rest, white sandy beaches stretch around numerous coves. The prevailing winds in summer are north and north-west. There are two harbors on the island: bays Ammos in the south and Fyki in the north. At a distance of 800 m to the east there is a small old fishing port (Avlakia), where yachts were previously not recommended to enter, since the entrance to it is closed by rocks, and the depths are variable (Fig. 5).

Currently, it has been completely reconstructed and can accommodate up to a dozen vessels, for which terminals have been built. When entering the line between Ammos Bay and Avlakia Harbor, reefs that are not marked by buoys protrude from the water. The origin of the island's name is unknown, but it is referred to as "Othronoi" or "Thoronos" in the writings of Greek authors of the third to tenth centuries. At one time the island was called "Fidonisi" (Snake island). The Italian name Fano has been known since the twelfth century and comes from the ancient Greek word "phanos" – "torch," indicating that a lighthouse was installed on the island. In Greek mythology, Othoni is associated with the heroes who participated in the Trojan War. One of them, Elephenor, the king of Avantes from Euboea Island, found

Fig. 5 Othoni. (Source: https://atcorfu.com/wp-content/uploads/2015/04/othonoi-port.jpg)

refuge here, but was expelled from there by a giant serpent. Another participant in the war, the famous king of Ithaca, Odysseus, after a shipwreck, ended up on the island of Ogygia and was held there by the nymph Calypso, who was in love with him. Ancient authors put the island of Ogygia in different places, and the version that it (mentioned in the "Odyssey" by the Greek poet Homer) was the island of Othoni itself appeared in the work of the Byzantine historian Procopius of Caesarea (sixth century). Othoni was deserted when in 551 the Ostrogoths, who sailed here from Italy in 300 ships, captured the neighboring island of Kerkyra. However, by the middle of the twelfth century, when the island called Fano was mentioned by the Muslim geographer Al-Idrisi, who lived in Sicily, it was already inhabited. In the middle of the fourteenth century Othoni, along with other small islands, was granted as a fief to Theodore Skaliti (from the island of Kerkyra) by the Neapolitan king Charles III. In the fourteenth to sixteenth centuries Othoni was often attacked by pirates: in 1537 the island's settlements were ravaged by the Algerian pirate and Turkish admiral Hayreddin Barbarossa. Only after the defeat of the Turkish fleet in the battle of Lepanto (1571), new inhabitants came on Othoni, which fell under the rule of Venice. They were from Ioannina and Parga (Epirus) and from the island of Paxos. They eventually colonized the neighboring islands of Ereikoussa and Mathraki. After the capture of Venice by Napoleon in 1897 Othoni, along with other Ionian Islands, became part of the French Republic, and in the period from 1800 to 1807 – to the Septinsular Republic, which was a protectorate of the Ottoman and Russian empires. In 1807, the island became part of the Illyrian provinces of the French Empire, but already in 1809 it came under the rule of England, and in 1815 became part of the United States of the Ionian Islands. In 1864, all the Ionian Islands were given by England to the Greek king George I. During World War I, in the Othoni region, the line of the blockade of the Strait of Otranto was passed by the ships of the Entente countries, which tried not to let the Austro-Hungarian fleet into the Mediterranean Sea. Near the port of Ammos, there is a neoclassical primary school building (1912) – now closed, Holy Trinity Church (seventeenth century) and a cemetery. To the northwest of Ammos there is the village of Dafni. There is the Church of the Ascension (XVII century) here, which was part of the monastery. To the southwest of Dafni lies the village of Chorio with the Saint George church (1864). A large stone white cross was erected near it on the Kalodiki hill in 1833 in memory of the devastation of the island and the massacre of the inhabitants, perpetrated in 1537 by the Algerian pirate and Turkish admiral Hayreddin Barbarossa. In 300 meters from the village there is a rock, which the locals call "tu Francesu to Koukouli" ("Frenchman's hood") – there is a local legend that there was a grave of French soldiers of the Napoleonic era. On the northeastern tip of the island, on a 100 meter high mountain, where a Venetian fortress once stood (its ruins have been preserved), a 13 meter high lighthouse was built in 1872. A beam of beacon light, sent every 10 seconds, is visible at a distance of almost 40 km. During World War II, the lighthouse was blown up by the Germans and rebuilt only after its end. Calypso's Cave is only accessible from the seaside on the western side of the island. According to local legend, the nymph Calypso lived in this cave, who for several years unsuccessfully tried to keep Odysseus by her side. A cave has two entrances. Its length is 100 m, width is 10–15 m, and height is 20 m. The basis of the island's economy is tourism. Agriculture is limited by the cultivation of vegetables and fruit trees (olives, figs). Fishing and olive oil production are well developed. There is a regular sea connection from the island of Kerkyra, as well as with the island of Ereikousa. Othoni is a traditional stopping place for migratory birds from Libya to Europe, especially common turtle doves. Hoopoes and judcocks can be met here during the winter months. There are also birds of prey – peregrine falcons and eagles. Among the marine life, bottlenose dolphins, some species of sharks, in particular, the white shark, and Mediterranean monk seal in the area of Calypso Cave are quite common.

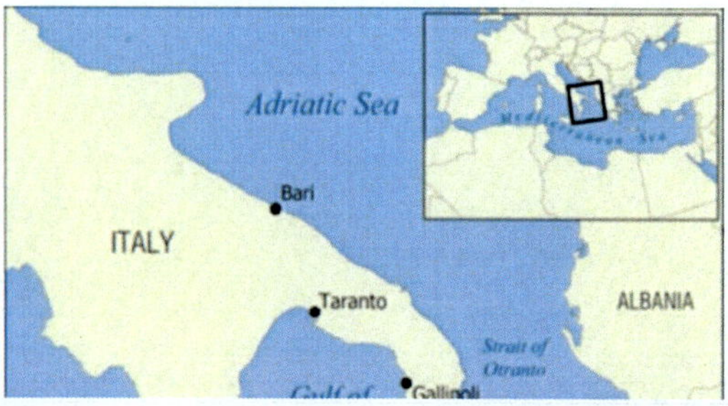

Fig. 6 The Strait of Otranto. (Source: http://1.bp.blogspot.com/_R82B80oQcsk/THxLjJM-hxI/AAAAAAAACVg/13WWrGW6JOQ/w1200-h630-p-k-no-nu/Gulf_of_Taranto_map.png)

Otranto

Otranto – (1) The strait between the Apennine and Balkan Peninsulas, between the coasts of Italy and Albania, connects the Adriatic and the Ionian Seas. Western shore belongs to Italy, and the eastern one to Albania. Width at the narrowest point is 72 km and the depth is approximately 740 m (maximum depth is 850 m). Named for the city of Otranto in Italy. During World War I, the strait was of strategic importance. The Allied naval forces – Italy, France, and the United Kingdom – blocked the strait (Otranto barrage), preventing the Mediterranean Sea, where the naval theater of military operations was located.

(2) The threshold is situated on the border of the Adriatic and the Ionian Seas, the depth of the threshold is about 800 m. To the north of the threshold there is the South Adriatic, which is an almost circular depression with a depth of 1230 m. South Adriatic is one of the areas of deep waters forming in the Eastern Mediterranean. the Austro-Hungarian Fleet from entering (Fig. 6)

"Otranto Barrage"

"Otranto Barrage" – the Franco-Italian net defense in the Strait of Otranto, between Fano (Italy) and Corfu (Greece) during World War I in 1918, which included minefields and a patrol of light naval forces. The main author of the project is Captain third Rank de Quillacq. Barrage installation lasted for 5 months. The length of the barrier is 66 km; it was divided into links, approximately 300 m each. The nets were 50 m high; their upper edge was immersed in water for 10 m. They were equipped with the mines of the de Quillacq system and were supported by buoys. Anchors and concrete blocks kept the system in such a way that even in the central part of the channel, where the depths reached almost 900 m, the barrage showed great stability at great heave of the sea.

Ottoman–Venetian wars

Ottoman–Venetian wars (1463–1718) (*Veneta-Turkey wars 1463–1718*) – seven wars between Venice and Turkey. After the capture of Constantinople by the Turks in 1453, Venice tried to establish permanent trade relations with them, however was forced to enter into a struggle, defending its possessions in the eastern Mediterranean. In seven wars with Turkey (1463–1479, 1499–1503, 1537–1540, 1570–1573, 1645–1669, 1684–1699, and 1714–1718), during which Venice entered into alliances with other European powers, it gradually lost almost all of its possessions – Cyprus, Crete and other islands, the coast of Morea (Peloponnese), retaining only the Dalmatian coast, several points in Albania and Corfu Island. The wars with Turkey have undermined the commercial and political power of Venice.

Oxeia

Oxeia (*Oxia*) – an island located at the mouth of the Acheloos River. It is private, bought by the Emir of Qatar, located east of Ithaca and separated from the mainland by a deep-sea strait. It is part of the Echinades group of islands. It looks like two islands. The area is about 4.2 km^2. It stretches 5.9 km in length, 1.5 km wide in the north, and 0.7 km wide in the south. Dense pine forests cover the island. Sheer cliffs surround the coast of the island on all sides. The island has three perfectly shaped, large, cozy bays that create natural harbors on the island. The northern part of the island is quite mountainous, its height is 421 m, and much higher than the southern one. It is densely covered with shrubs and pines. The pure marble of high quality is found on the island. There are a lot of fish in the surrounding waters. Local fishermen have built arched nets in the large bays of the island. The nets can be removed or used further. The island is uninhabited and has not yet been affected by civilization. There is only an abandoned, picturesque lighthouse on the island.

Oysters (Ostreidae)

Oysters (lat. Ostreidae) – an Ostreoidea family of marine bivalved mollusks. They have a very specific asymmetric shelve of irregular shape. One of the most popular among the commercial groups of marine invertebrates – thick-walled and inequivalve. It consists of a larger convex (mostly left) section, which becomes attached to various underwater objects, and a smaller, flatter and thinner, free section, forming a kind of cover. The apex of the sections is straight, on the right usually larger than on the left; the scarfwise without spikes, the retinaculum connecting the two sections, is located at the scarfwise on the inside. An immantle is attached to both shell sections (separating the shell). On the inner surface of the shell surface, there are visible imprints, that is, the points of attachment of one adductor muscle, with the help of this muscle both sections close together; however, this process has not yet been clarified by scientists. The gills of the oysters consist of two thin plates on each side of the body, covered (as well as the mantle) with ciliary hairs supporting a continuous flow of water around the body of the animal. Thanks to the action of all these ciliary hairs, the animal constantly receives fresh water rich in oxygen, as well as various food particles suspended in sea water, both dead and living, consisting of unicellulars (infusoria, algae), rotifers, and small larvae of various marine animals (coelenterate, worms, mollusks, etc.). The size is different; giant oysters can reach a length of 38 cm; however, European species usually have a shell length of 8–12 cm. The oysters are diclinous; under favorable conditions, one female can produce up to 500 million eggs per season. Spermatozoa flow with the waterflow into the mantle cavity of the female, and fertilized eggs develop in the back of the mantle cavity. After a few days, moving larvae enter the water; they swim for several days and then settle, having a well-developed leg. With the help of the leg, the young crawl, choosing a place for the final cementation, after which the leg begins to decrease and, no later than after 72 days, is reduced completely. They live both singly and in colonies, usually on solid grounds – stones, rocks, or mixed sandystony grounds at depths varying from 1 to 50–70 m. It is customary to distinguish coastal settlements and oyster banks. Like many coastal animals, these mollusks are able to tolerate some desalination; the minimum salinity at which they can exist is 12‰. Salinity level affects the growth rate of oysters and their taste qualities: oysters collected at salinity from 20 to 30‰ are considered the best, where there is a small and constant desalination by river water. In conditions of the salinity about 33–35, oysters grow well; however, their meat becomes hard. This specificity was well known to the ancient Romans, who kept oysters collected in the sea in small desalinated reservoirs On the settlements and banks, oysters sometimes live very densely, then their shells often stand vertically, with the abdominal edge upwards; sometimes they grow together in several pieces. During severe low tides, the shell colonies are sometimes exposed, so many species have the ability to survive long drains. Oysters are usually consumed live and raw, served on ice, and relished with lemon juice. Mollusks can also be fried, baked, boiled, and cooked in small amounts of water. Oysters have high nutritional value. The presence of fine taste of oysters is due to the specific chemical composition, which also has a tonic effect on the nervous system. Oyster meat contains fat, protein, glycogen carbohydrate, nicotinic acid, and vitamins of group B. Among other things, oysters contain the following minerals copper, iron, calcium, zinc, phosphorus, and iodine. To provide the body with a daily dose of copper and iron, it is enough to eat six oysters. Oyster meat is

Fig. 7 Oysters. (Source: https://cepia.ru/images/u/pages/1616/ustrica-3.jpg)

saturated with a complex of fatty acids, especially useful for humans, among which is Omega-3. As a result of research conducted by American scientists, in oysters were found ceramides – fats that can fight breast cancer. Ceramides block and inhibit the growth of cancer cells. American and Italian scientists have found that raw oysters, mussels, and some other types of shellfish contain unique amino acids that affect the active production of sex hormones. Zinc contained in oysters is necessary for the synthesis of testosterone in the female and male body (Fig. 7).